AF556934

By Design

By Design

The story of Crown Equipment Corporation

by Pat McNees

ISBN: 1-882203-15-1

Orange Frazer Press, Inc.
Box 214, 37 1/2 West Main Street
Wilmington, Ohio 45177

Library of Congress Cataloging-in-Publication Data
McNees, Pat.
By design : the story of Crown Equipment Corporation / by Pat McNees.
p. cm.
Includes index.
ISBN 1-882203-15-1
1. Crown Equipment Corporation--History. I. Crown Equipment Corporation. II. Title.
TJ130.C76M36 1997
338.7'62--dc21 97-25588
CIP

Printed in Canada

Cover: Graphica
The photograph of Orson Welles on page six is by permission of Photofest, 22 West 23rd Street, New York, New York, 10010
Photographs on pages *v, x, xii, xvi* and 82, 109, and 118 by Eric Albrecht

For Jim Dicke Sr., and the rest of the Twenty-Five Year Club

Acknowledgments

This story about Crown Equipment Corporation came into being because, in the course of researching another book, I became fascinated with the story of how Crown — a company known for making heat regulators and television antenna rotators — had cleverly and quietly slipped into the lift truck industry and become a major competitor. Not only was the story that unfolded more interesting than I expected, it was unexpectedly heartwarming. Many of the people I interviewed had worked at Crown for most or all of their working life, and twice when I was interviewing retired employees I was surprised to see tears in their eyes — tears of gratitude and nostalgia for a company that, in starting nowhere and working its way to the top, had given them a chance to do things they'd never dreamed they could do. This company was not only interesting and admirable; it was loveable. It serves as an excellent model for both entrepreneurship and management. I wish there were more companies like it.

I am grateful to many people for interviews, including Tom Bidwell, Roger Bornhorst, John Caywood, Dennis Dicke, Eilleen Dicke, Jim Dicke Sr., Jim Dicke II, Paul Dicke, Andy Doenges, Karen Eckert, Lawrence Egbert, David Goldzwig (of Joyce-Cridland), Duane Hegemier, Arnie Heitkamp, Verlin Hirschfeld, Carol Jones, Don Luebrecht, Betty Marshall, Bob Marshall, Lois Moeller, Jim Moran, Jerry Pulskamp, Al Rose, Irene Schlueter, Wayne Schroer, Leo Schulte, Harold Stammen, Kathy Topp, Jim Uetrecht, Tony Van der Straaten, Dick Watercutter, Warren Webster Sr., Wilbert Will, Mose Zimpfer, and Charles Zwiebel. I am especially indebted to Tom Bidwell, Paul Dicke, Arnie Heitkamp, Jerry Pulskamp, Wilbert Will, and (bless his heart) the painstaking Harold Stammen for endlessly patient technical explanations. Whenever a detail or explanation eluded me, Dee Fledderjohn managed to hunt down the right person and get them to stop what they were doing long enough to answer my questions. Thanks in countless ways to Bill Klosterman, Pam Bergman, Dave Helmstetter, Mike Spicer, and Sue Luebke, in Crown's marketing communications department and to Carol Jones, Julie Ahlers, and Laura Anderson, in Crown's executive office. Thanks most of all to Jim Dicke Sr. and Jim II for being so open about their business.

Thanks to Kitty Kelley and the writer-referral service of the New York-based American Society of Journalists and Authors (ASJA) for putting me on Crown's radar screen. Thanks to Sue McNees for transcriptions above and beyond the call of duty and to Jim McCready for correcting a number of technical errors. Many thanks to Lori Schubert for the manuscript's final edit, to Marcy Hawley of Orange Frazer Press for coordinating production, and to Brooke Wenstrup and John Baskin for patiently and intelligently designing the book and shaping the final product.

Many of the insights about lift truck design and product development were influenced by a 1991 case study ("Crown Equipment Corporation: Design Services Strategy") that was co-sponsored by the Harvard Business School, partially funded by the Design Arts Program of the National Endowment for the Arts, and published by the Design Management Institute (29 Temple Place, Boston, Massachusetts 02111). The case was prepared by Karen F. Freeze, director of research of the Design Management Institute, and Gary Pisano of the Harvard Business School as the basis for class discussion. Excerpts from the case study are reprinted here by permission of the Design Management Institute. For the chapter on New Bremen, I drew on material from an article by Jane Ware that appeared in *Ohio* Magazine. In other chapters, I used material from my earlier biography of Warren Webster Sr., *An American Biography: An Industrialist Remembers the Twentieth Century* (Farragut Publishing Company, 1995).

Pat McNees

Bethesda, Maryland, April 1997

Alice Barhorst at work in Plant 1's rotator assembly

Contents

Introduction

Today Crown Equipment Corporation dominates the market for narrow-aisle lift equipment, but in 1945, when Crown Controls was launched in New Bremen, Ohio, no one would have predicted that this would be a significant firm in any field. Shortly after World War II, the firm began manufacturing and selling temperature controls for coal-burning furnaces, a product it marketed only briefly, as home builders began turning to gas heat and the market for coal all but disappeared.

In a search for new products and new markets, Crown diversified in several directions beginning in 1949. First, in response to an emerging market for television, the firm began producing television antenna rotators, a product for which its sales have remained stable, despite enormous changes in communications.

From the 1950s through the 1970s, as much as twenty percent of the firm's business was contract manufacturing and repair services for the U.S. military and for private corporations as varied as Baldwin Pianos and IBM. In addition to supporting Crown's overhead and keeping its factories busy, the contract work provided Crown with precision machinery, superb training, greatly increased know-how in important areas, and a reputation for quality work. Even a brief period of producing precision-manufactured but low-profit novelty products in the mid-fifties helped Crown develop first-class manufacturing skills and valuable experience and, ultimately, the reputation that went with them.

The contract work together with sales of antenna rotators subsidized the development of what would become Crown's most important product line. By 1959, three years after shipping its first unit, Crown had become a niche producer of small manual fork lifts. For most of its history, Crown was not considered a significant competitor in the lift truck industry. But the firm grew and developed, slowly but steadily, over forty years in the material handling industry, adding products a year at a time until it had a complete line of heavy-duty electric lift trucks. In 1988, Crown Controls became Crown Equipment Corporation. It is still a privately held, family-owned firm.

In 1995, fifty years after it opened, Crown owned factories in five countries and branches in eight, and employed thousands of people worldwide. It had become the largest manufacturer of electric material handling equipment in the world and the leader in the narrow-aisle lift truck market. Today Crown can sell you lift trucks with lifting capacities from 1000 to 8000 pounds and lifting heights from five inches to 45 feet. By 1995, the only thing Crown didn't make was gasoline-powered material handling equipment.

This is the remarkable story of the early decades of Crown's development. It illustrates that the shortest distance between two points is not necessarily a straight line, and that one key to business success and longevity is to make the most of the material and people you have on hand. Old-timers recall Crown being more of a family operation than an eight-to-five job, with everybody working together on many projects, well up into the material handling period. Even the engineers weren't stuck at a desk all day: When it was time to haul bricks, bale hay, or carry in a load of steel, all of the men were expected to drop what they were doing and pitch in. Few workers were stuck doing the same thing over and over — the sort of work that makes an eight-hour day seem to last forever — because most workers had multiple responsibilities. Teamwork and flexibility were particularly important in the early years when Crown was working with used equipment that required a good deal of patience and ingenuity. Gradually the firm acquired the skill and machinery to manufacture components internally, and today it manufactures roughly 85 percent of the components in its trucks, a remarkable figure in a world where outsourcing is increasingly common.

Although tools, machinery, equipment, and vehicles play a role in this tale, so do hard work, perseverance, lucky breaks, and sound business decisions. But it is mainly a story about people — about small-town people working together to develop a world-class product for an international market. It's the story of a company and a group of people who grew up together.

By Design

The early years *Salesman Carl Dicke (center) started his career whitewashing walls but in the late 1920s went into business with his two brothers (not shown here). In 1945, they launched Crown Controls in the old Rabe Hardware store (opposite page). In an energy market that would soon shift to gas, their temperature controls for coal-burning furnaces had a limited lifespan.*

It All Started *with a* Heat Regulator

(1945–51)

Wm. Rabe's Store and Office.

1.

Crown's origins were inauspicious: The firm began operating in 1945 in a building not much larger than a garage. The seeds for the company had been planted much earlier, however. In 1917, the Remington Arms Company hired Allen Dicke (the name rhymes with Mickey) of New Bremen, Ohio, as a patent attorney. Remington, an arms manufacturer, had just begun to diversify into business machines. Allen worked first in the firm's Bridgeport, Connecticut office and later moved to the Ilion, New York office. For many years he was involved in litigation with National Cash Register (NCR), a Dayton, Ohio firm with a major share of the market in business machines. In 1929 Allen became a vice president of Remington Arms; in 1932 he started his own practice in New York. Meanwhile, his younger brother, Carl, had worked his way from white washing basement walls in Remington's Ilion office to a position in sales.

In 1927 Carl and Allen decided to go into business for themselves with their older brother Oscar, an inventor. Oscar held more than a hundred patents on various clocks and timing devices associated with railway signaling. Allen had invented an improved thermostat for regulating room heat that, in combination with other standard control devices, could be used to regulate a hand-fired coal-burning furnace. Carl and Allen also developed a control (to be located in the bathroom) that lit a gas-fired "one-arm" water heater (generally located in the basement). The Sav-u-time, as it was called, showed a pilot light when the heater was operating. They sold 200 of the devices in the New Jersey builders' market.

Together the three brothers — an inventor, a patent attorney, and a salesman — started the Pioneer Heat Regulator Company in Bloomfield, New Jersey, with Oscar participating only financially. The heat regulator sold well in the 1920s, but events conspired to send sales tumbling as the Depression hit: At a time when few new homes were being built, John L. Lewis organized the coal miners, coal prices rose, and natural gas became competitive. In 1931 the brothers moved the plant from New Jersey to New Bremen, Ohio. New Bremen was a farm town of 1,500 people where the Dickes had been raised (in a family of six children) and had farm interests. The family roots went back to New Bremen's earliest days.

Oscar and Allen

Carl's original partners were his brothers Oscar (left), an inventor, and Allen, a patent attorney. Oscar became wealthy from royalties on clock mechanisms he'd invented and held patents for and on new technologies in which he'd invested. Paying a dollar a share for stocks such as IBM and Xerox, he died a wealthy man, partly, says the family, because he "still had the first nickle he ever made."

After struggling for a year to keep the firm alive, in 1932 they sold the business to Master Electric. As a condition of purchase, Master Electric insisted that Carl move to Dayton to manage the firm's Heat Regulator and Small Appliances division. Carl did so until 1943 when, at the age of 41, he had his first heart attack. He recovered but he retired from Master Electric. After serving a brief stint as president of the R.C. Allen Co., a troubled Grand Rapids, Michigan firm, he remained in semi-retirement — but was bored. In 1945, toward the end of the war, Carl and Allen hatched the idea of starting a small business to market an improved version of their earlier heat regulator. When the thermostat called for an increase or decrease in heat, the damper would open or close, to hold the temperature in the house steady. In 1945, as Crown Controls, the tiny firm began marketing thermostats and temperature controls. Carl's son Jim Dicke, fresh from service in the Air Force, joined the company immediately. Jim began working in 1945 but didn't go on the payroll until January 1, 1946. At that point, Crown had only four employees: Carl Dicke, Jim Dicke, Bill Havemann (an accountant and general business manager), and Lois Gensler Moeller (a part-time accountant).

The assembly line

"Make it look like you know what you're doing," Carl Dicke told accountant Lois Moeller before a photographer took this picture of Crown's "assembly line." Left to right: Carl Dicke, William Havemann, Carman Hirschfeld, Jim Sr., Julitta Nedderman, and Lois Moeller. In 1945, the tiny firm had begun with only four employees.

Timing is everything in the development of a company. It is important that the right facility be available when a company is ready to start operation. Crown almost went into business in Fort Loramie, Ohio, where an aggressive group of bank and community leaders were encouraging new businesses and a local banker was ready to help the young company build a plant on a five-acre site. At the last minute the old Rabe Hardware Store at 124 South Washington Street became available in New Bremen. The Dickes chose the hardware store because the second-floor joists were heavy enough to hold the machinery Crown would need for manufacturing. For about eight months, Crown leased the back two-thirds of the downstairs to a creamery for storage and, from a one-room base, began servicing a handful of customers, notably Blue Coal, the largest home supplier of coal in the industry.

Plant 1, as it would later be called, was really two buildings, for which the Dickes had swapped a coal yard owned by F. T. Kamman, Carl's father-in-law. The northern half of Plant 1 was the hardware store, 2900 square feet, which Crown soon took over altogether. The southern half, next door, was Sophie Kellermeyer's house (not to be confused with the Ekermeyer house north of them — German names abound in New Bremen). By 1948, Crown was using both floors of the hardware half of the plant. Downstairs were the office, the first toolroom ("the lab"), and the shipping and receiving department. Upstairs was devoted to assembly.

In the early years, Crown manufactured only the thermostat — that is, five ladies working side by side along a ten-foot section of conveyor belt did so. Crown bought its damper motors from White Manufacturing and its transformers from Chicago Transformer, assembled them with the thermostats, and sent most of the finished heat regulators to New York, to the D L & W Coal Company (known popularly as "Blue Coal") to be used on coal-fired furnaces. Blue Coal was a division of the Delaware, Lackawanna, & Western Railroad System.

At the time, coal dealers customarily sold and serviced coal furnaces, just as some gas companies today sell and service gas furnaces. Blue Coal mined anthracite coal, the hardest, best-burning, top-quality coal, preferred for home use. This coal burned with a pale-blue flame, which the company highlighted by spraying the coal with blue dye, to show its superiority and distinguish it from other coal. For many years, Blue Coal was a sponsor of the radio program, "The Shadow." During the intermission, an authoritative-sounding man would come on to offer tips on how to fire your furnace and how to bank it for the night. In the middle of the program, when Lamont Cranston, The Shadow, was about to catch the villain, the announcer would interrupt to say, "In addition, you should use a Blue Coal Tempmaster heat regulator system to further improve the efficiency of your operation." Those were the heat regulators Crown made in New Bremen.

The Shadow knows

Between segments of the popular radio program, "The Shadow," whose hero was played by Orson Welles, the sponsor would interrupt to market the Blue Coal Tempmaster heat regulator, made in New Bremen.

Only six months after Crown was launched, Master Electric called Carl and told him that Crown had taken all their customers, that they hadn't had a new order for thermostats for thirty days, and that they'd decided to get out of the heat regulator business. Master Electric offered to sell Crown all of its tooling for $85,000 if Crown would take over its warranty business. In 1947, Crown incorporated and bought Pioneer back from Master Electric, promising to honor existing Master Electric orders and to handle all warranty and service work. They hauled ten semi-truckloads of tooling goods and inventory to New Bremen and moved them into the old Rabe Hardware Store building.

The Dickes had not wanted to be a manufacturing business, but now they were, and they had all the problems of a new manufacturing firm with no money and little manufacturing know-how. To make things worse, Master Electric had changed a bearing bracket design. Instead of a fixed bearing bracket, they had used a piece of blue steel spring, so that if you dropped the box or jarred the regulator in shipping, the bearing would lock in position and the regulator would not function. Every one of those heat regulators had to be rebuilt; they were coming back literally by the boxcar-full to New Bremen. After Crown identified the problem, the president of Blue Coal said to his people, "Frankly, the reason you fellows got bad material from the Dickes is that you didn't pay them enough money; they couldn't make a decent product."

"So," says Jim, "They paid us $1.50 more apiece. They were super people. We would produce two and repair one. That gave us the cash flow to keep going. That's how we stayed in business."

Initially Crown assembled the heat regulators after buying two-thirds of the components. After buying Pioneer back, it began manufacturing more parts and strengthening its manufacturing capabilities. Harry Gilbert, who had worked for Pioneer under Master Electric, was asked to set up production in New Bremen so that Crown could make its own transformers. The same year, Crown began making its own damper motor (the C1244). The people in Crown's small engineering section worked hard to improve that motor. They started with a small transformer, then worked their way up, developing a power-failure unit that would reverse the motor when there was a short circuit or a power failure, and close the furnace damper, so the house wouldn't burn down.

Crown manufactured one thermostat for Blue Coal and another for other coal and coke companies to sell with customized insignia. Eventually the firm began expanding sales by making specialized models of its main product and by doing contract work related in some way to what it was already manufacturing. Among other specialized units, Crown manufactured a modulating control for school ventilating systems. The state of Ohio required that a certain proportion of the air in school classrooms be fresh, unheated air from the outside. As the temperature changed, the proportions of heated and fresh air changed, so Crown devised a system that would modulate the damper system and maintain the appropriate proportions.

Crown also developed and manufactured two overheat thermostats, one for hot air furnaces and another for hot water heaters. One of these was used by Blue Coal for its "Bucket a Day" water heater. This appliance stood next to the furnace in the basement; every day you put a bucket of anthracite coal into it that would heat water for the house all day. The coal-fired water heater was useful in areas that did not have access to gas.

Another product Crown manufactured for Blue Coal was an Ash Pit Spray Kit, to hold down coal dust. Allen Dicke's son, Paul, who had joined the firm in April 1947, explains, "In the days of coal-fired furnaces, after you burned the coal it ended up as ashes in the pit. Messing with ashes when you are cleaning up the ash pit floods the basement with dust, so Crown devised a spray arrangement to fasten beneath the grate. There was a little brass nozzle cut to throw a fine water spray. Before you started cleaning the ashes you would turn this thing on just a little bit and as you shoveled your ashes into the garbage pail or ash barrel you didn't get dust all over the place."

By 1949, Crown had 35 employees. The Dickes were still experiencing the typical struggle of a startup business: collecting enough money so that cash was flowing in as well as out. "Carl would ask me to work an extra evening once in a while," recalls Lois Moeller, "but rather than pay me overtime he would bring home a pair of silk hose for me from his next trip to Dayton. That was my overtime pay. Or if he had to meet a client in Dayton or somewhere, he would take me along in case he needed some notes taken, for which he would buy my lunch."

The Dickes got advice from Warren Webster (Eilleen Dicke's father and Jim's father-in-law), an experienced manufacturer who co-owned the Joyce-Cridland Company. Joyce-Cridland was a Dayton firm that manufactured hydraulic lifts for the service and repair of autos and other vehicles. Jim had been telling Webster about Crown's problems with cash flow, so one day Webster showed Jim and Carl a new budget system he had set up at Joyce. The Dickes didn't have much capital, but Webster showed them how to find a little money in the budget for capital improvements, and they went back to New Bremen and set up a budget system. Webster's advice was so helpful that they invited him to be on their board of directors, a position he held until his death in 1994.

Sounding board

Eilleen Dicke's father, Warren Webster, shown here with his wife, Mary, gave Jim Sr. invaluable advice. In the early days, Jim was on the phone almost daily with his father-in-law, an experienced Dayton manufacturer.

Meanwhile, Crown was having management problems. Carl was an excellent salesman, and the manager on the spot in charge of daily operations. He'd never run a whole company but he had a pretty good practical sense of how a business should be run, and he had Jim to help him. Allen was a New York attorney, very structured, and careful with his money. "Carl and Allen had been very close earlier in their lives," explains Jim II, Jim Sr.'s son and Crown's current president, "but by training and temperament they were very different. Trained as a lawyer, Allen was methodical and thought of business the way he thought of the law: as a chess board, with moves to be made and rules to be followed. Carl, on the other hand, had been primarily a marketing and sales whiz throughout his career, and like most salesmen, believed that things like careful accounting and cash flow analysis were a 'bean-counter' function. He probably did not recognize the difficulty and challenges that come even with a successful and expanding business, the cash needs that can force a company to the wall even in the face of strong orders." Carl and Allen spent a good deal of time writing letters back and forth to each other about how to run the business, and friction developed. They argued with more and more frequency.

"One night Uncle Carl came to visit," says Paul Dicke, "and they sat in my Dad's study. At the far end of it was a drafting table where I was working, designing a damper motor. They started talking about how Carl wanted to do this and Dad wanted to do that (almost the opposite). My dad said, 'By what authority do you make this statement, Carl?' My Uncle Carl stuck his thumbs in his vest and said, 'Why, my broad experience in sales.' My father said, 'Well, disregarding that, what . . .' They were at each other's throats. I don't think either one of them was really up on modern methods of business operations."

As the 1940s came to a close, it became clear that Carl

and Allen should not stay in business together. Allen told his son, "Paul, you know I want to be friends with my brother, and ever since we've gotten together in this business we haven't been able to see eye-to-eye on anything. It's more important that we live as brothers than as combatants. I can take the money I have in this business and put it in the Chemical Exchange Bank in New York and make more money than I'm making now. What would you think if I got out of it? What would you do?" And Paul said, "I think I'll stay on." Sometime later Jim Sr. came to Paul and said, "If you do stay on, you'll not regret it." Apparently it was a wise decision. "Talking to Jim through the years," recalls Paul, "I always said, 'I don't know of many days that I got up and said to myself, "Oh gee, I've got to go to work."' Work at Crown always was fun and a challenge. We were so small you could involve yourself in almost anything."

Life of a salesman

An excellent salesman, Carl Dicke (at left) had never run a company but he had a pretty good practical sense of how a business should be run, and he had Jim to help him. But in 1952, at the age of 50, Carl died of a heart attack, leaving 30-year-old Jim to run the fledgling firm.

Finally, Carl and Allen agreed that one brother would buy the other out. One brother would set the value of the business and the other would decide whether at that value he wanted to buy or sell. They flipped a coin: Carl set the value, and Allen said that at that value he would like to buy it, but he would like to hire Jim Sr. to stay and run it. Jim wasn't interested in working for Allen, though, and Allen didn't want to run it himself. So in 1950, Jim and Carl bought out Allen's share of the business. Carl put up nearly everything he owned; Jim pledged "sweat equity," his willingness to work for almost no money. Allen and Carl shared ownership of a farm, so Carl exchanged his interest in the farm for Allen's interest in Crown. Allen retired from his New York law partnership that year and settled down on his New Bremen farm. Allen's grandchildren reported regularly to their cousin, Jim II, that Carl and Jim were likely to go bankrupt soon, at which point Allen would no doubt buy the assets out of bankruptcy. Allen was surprised that the business survived and did as well as it did, because Crown had so little working capital. He remained convinced that Warren Webster was a stockholder, a silent investor in the company, but Webster never was. He had more confidence than Allen did that Jim and Carl could make a success of the firm.

By the time Jim Sr. and Carl bought out Allen's interest in the firm, Jim was effectively managing Crown. Carl was still semi-retired and taking things easy because of his bad heart. Crown had installed a one-man counterbalanced rope pulley elevator so Carl could get to the second floor of Plant 1 without walking up the steps. The elevator worked fine when he rode alone. When he brought his grandson, Jim II, on the elevator with him, the extra weight of Jim's body forced him to pull extra hard on the rope, and he had to stop two or three times to catch his breath before they reached the second floor. A warm, good-humored man, Carl sometimes picked Jim II up and took him to Crown meetings. The boy enjoyed sitting in on the meetings, although he eventually suspected that he was his grandfather's ticket to leave them early. Carl would say, "Gotta go now. Have to get the boy home, you know."

In December 1952, Carl Dicke died unexpectedly at the age of 50. With his passing, the young firm lost years of experience. Jim was now on his own running the company, at the age of 30 and with only seven years' experience in business. "When Carl died," says Lois Moeller, "Jim got thrown into (the presidency) so quickly he hardly had time to get his feet wet, but he showed authority — he showed that he could handle things. He had an aura about him, just the way he carried himself, that made you feel that, yes, he is your boss."

The ties that bind

Through Dicke family gossip, Jim Sr. knew that Uncle Allen expected the fledgling firm to go bankrupt—and was surprised when it didn't. Allen sold his interest in Crown to Carl and Jim in 1950 but at family gatherings felt free to express his opinions (for example, that Jim should give relatives in the firm better raises). The Dickes shown at this family gathering are, left to right (standing), Allen Jr., Carl, and Allen Sr., and (seated) Irene (Carl's wife), Eleanor (Allen Jr.'s wife), and Amber (Allen's wife).

In a small and struggling company, there is often a thin line between home and work. In the company's early days, Jim did a lot of traveling to seek new customers. When he was home, after having dinner with the family and sending his sons Jim II and Dane to bed, he would go back to the office to do more work. "I don't think New Bremen was a particularly easy place for Mom [Eilleen] and Dad at first," says Jim II. "Although Dad's family had lived in New Bremen for more than a hundred years, he himself was seen as something of an outsider, having been raised primarily in Dayton, Ohio. Even more so than Dad, Mother was a Daytonian and did not have a large circle of local friends. She did her best to make friends and become part of the community, but the opportunities weren't as great as they would be today. She was involved with a youth group called the Child Conservation League and belonged to a bridge club, but many times we spent weekends in Dayton at the Webster home, where, of course, Dad and Grandpa Webster discussed business most of the weekend. Mom and Grandma would disappear for the day to go shopping or run other errands. I usually stayed with Dad and Grandpa while they talked about Grandpa's latest frustration with some business problem or Dad's concern about an efficient production schedule."

Jim Sr.'s father-in-law was his mentor, a beloved and valued sounding board over the years. If Jim had a problem, Webster usually had a story or two to illustrate some point that might help solve it. He advised Jim not to use a small local accountant, for example, but to hire a certified public accountant, who had to have had enough education to pass a state examination. He recommended that Crown hire a large law firm because a large firm could afford a varied staff with expertise in many areas. At one point Jim was offered a good salary to run another company and was tempted to accept. Webster, who knew Jim wouldn't be satisfied working for somebody else, persuaded him to forget the whole thing. "His strong point," says Jim, "is that he never really interfered but he always helped."

Through the heat regulator business, Jim had become well-acquainted with John "Pop" Shipherd, the president of Blue Coal. In the mid-fifties, Shipherd told Jim about a problem and asked Jim for help. The Hershey Motorstokor Company, which made stokers for anthracite-burning furnaces, was going out of business. The few anthracite mines still being worked were on the east coast, and Hershey was one of the last major manufacturers of stokers for that kind of coal. Shipherd told Jim his firm didn't "know anything about manufacturing. We want you to take it over and we'll finance it. You make stokers for us, and we'll sell them." Blue Coal didn't know how many stokers they would sell, but with Webster's help Jim worked up some "IF/THEN" projections and took the deal on. He bought Hershey and all its tooling from the Fuller Machine Company, of which Hershey was a division, and changed the firm's name to Hershey Motor Stoker, Inc. For two and a half years Jim ran two companies at the same time, going up to Blue Coal, in Lancaster, Pennsylvania, for two days every two or three weeks, "until finally Blue Coal decided it wanted out of the arrangement," says Jim. In 1957, when Hershey Motor Stoker Inc. went out of business, says Jim, "Blue Coal bought us out and treated us very well."

In about 1959, the Blue Coal Motor Stoker Company also went out of business, because the market had changed. In the 1920s and 1930s, big oil and gas fields had been discovered in Texas and Oklahoma and the technology was developed to pipe gas from Texas to heat homes in other states. The market for coal stokers began declining; Blue Coal had a good product but few customers to buy it. Later, during World War II, only soft coal was available to civilians. Soft coal was smoky, messy, and unreliable, and after the war most people either went to gas or converted their old coal furnace to an oil burner. Where gas was available, new buildings normally used gas heat.

"The people in the coal industry totally misread what was going on," says Jim II. "They thought that if they could just update their product, business would return to normal. And as everyone knows, it didn't." By 1951, Crown had begun exploring new ways to produce revenues. Clearly, demand for its main product, temperature controls for coal-burning furnaces, was declining and the firm needed a new product. ❍

Small-town strength *Part of Crown's strength lay in its small-town base. Most Crown employees grew up on farms in and around New Bremen, and the firm benefited from that rural work ethic (eight hours' work for eight hours' pay) and a stable, mechanically adept, problem-solving workforce accustomed to teamwork. Inset photo is the turn-of-the-century farm of Henry Dicke, Carl Dicke's father.*

A Strong Base *in a* Solid Farm Community

2.

The story of Crown's development is in no way the story of a one-man band: Many people figure prominently in the company's success. "Crown's top management" say the authors of a Design Management Institute case study of Crown co-sponsored by the Harvard Business School, "believed that people were the real key to the company's success. Crown's talent came not from MBA programs or prestigious engineering schools but from the villages and farms of the surrounding countryside. Most of Crown's senior managers had been with the company since its beginnings or shortly thereafter, and most of its employees had spent all of their working lives at Crown. Many had grown up on family farms. As Don Luebrecht, senior vice president and former project engineer (and still a practicing farmer), says, 'Farming helps keep your sense — as a user and a fixer — of what's real and what's not.'

"Many employees had left the area for their education, but then returned to be near their families, or to have 'a good place to raise kids,'" says the case study. "The rural background of the workforce resulted in a stability, a distinctive work ethic and an individualism that contributed substantially to Crown's culture and the quality of its products."

"Looking back," says Tom Bidwell, Crown's recently retired executive vice president and general manager of material handling, "we're fortunate that we started up here in a rural area of Ohio. At first we thought it was a disadvantage that we weren't in a major metropolitan area, that it would be hard to get people. As it turned out we started where everybody else is going anyway. They're moving out of urban areas into the sunbelt and the rural areas, seeking that better work ethic. We grew up with this rural work ethic, eight hours' work for eight hours' pay. A very conservative community. But that alone doesn't do it. You still have to communicate. Everybody has to take pride in the job they're doing. It's never just the money."

Adding value

Bill Havemann, at left, working with Carl Dicke and Carman Hirschfeld, was Crown's first employee and head of accounting. Bill was an older man with good judgment whom Jim Sr. often consulted after hours while Bill planted flowers or worked on his lawn. In those days, employees were still paid in ten, five, and one dollar bills and a nickle-an-hour raise was a big deal.

"If Crown had located in a large city, chances are it would not be what it is today," concurs Tony Van der Straaten, who got involved early in Crown's international sales. "Because it is located in an area with a European heritage, where family values are still high, the turnover of the workforce is extremely low. As a result, our efficiency, productivity, and flexibility are generally higher than they would be in a big-city environment. In a city, the commitment of the workforce to management, and vice versa, would be much lower than it is in Auglaize County."

Although New Bremen was once connected by streetcar

(the Interurban) to St. Marys and Fort Loramie, New Bremen "was smack in the middle of farming country," says Jim II (Jim Sr.'s son and Crown's current president). "This probably had a lot to do with Crown's success. A lot of the employees have small farms." How does this affect Crown? "Well, first of all," explains Harold Stammen, manager of Crown's research and development and a long-time employee, "farm people tend to be mechanically inclined because they grow up with mechanical devices. They buy machinery because they know what machinery will do for them; they have an expectation of it. It's not like they're just turning their lights on. They have exposure not only to automobiles, but to farm tractors, combines, and all the other mechanical equipment. The other thing is that farm people know how to work. Truthfully, some people don't know how to work. They don't want to put in much effort. They don't persevere." Stammen goes on to point out that "this is a German town. Is there such a thing as a lazy German? My understanding of German characteristics is that this is almost a contradiction in terms. When I grew up we always had to be busy. Even as children my dad saw to it that we were kept busy. He would never let us stand with our hands in our pockets because that was a sign of laziness."

As Warren Webster explained, "Farm labor usually works out pretty well in manufacturing because farmers grow up realizing that if you don't do the work you can't harvest a crop. And farm kids grow up doing chores so they consider work as a part of life, not as an imposition." Farm people also understand process. They know that if they want a crop of potatoes they must do A, B, and C in the right sequence at the right time. If they want peanuts they must do X, Y, and Z, different steps on a different schedule. And farmers learn to avoid waste, to make use of everything — use a thresher, for example, to separate peanuts from peanut vines, then use the vines as feed for livestock. Saving money with peanut hay is a good precedent for creatively using factory resources." In fact, from the beginning, when Allen and Carl still owned both a farm and Crown in common, at harvest time Crown employees would often be told to leave the factory to bale hay for the farm. (Many employees also took off November 15th, the first day for hunting rabbits and pheasant; if they got their limit early, they would come back to work.)

"We are still hiring good, qualified people," says Tom Bidwell. "Most of our operations around the country are being run by people who started right down the street in the plant here and did a good job and became supervisors and worked their way up. These people in California and New York, wherever you go, are basically people from here. Good, honest, hard-working people you trust, who started in sales here in New Bremen and became managers of a branch. A lot of talent has come from this area."

"Morale was great at Crown," says Wilbert Will, "because Crown treated people right. I don't know of anybody who has a real nasty job here. Crown doesn't have many robots but they do have some robot welders who do the repetitive work in welding."

A lot of Crown's strength emerges from its base in the small community of New Bremen. "Many of our people here have grown up together," says Tom Bidwell. "We know each other. We trust and respect each other. And we have built on these positive relationships to help make Crown grow." This mutual respect and ability to work together as a team would be invaluable as Crown sidestepped into a totally new product line. ❍

Smooth transition *The tooling and technology used to manufacture an antenna rotator were similar to those used to manufacture temperature controls, so the transition from one product to another was easy—and there was plenty of work for women.*

A Train Ride Later: *the* Television Antenna Rotator

(1949-57)

3.

By the end of the 'forties, the heat regulator business had weakened, and Crown needed to find a new product, preferably one that would use the firm's new expertise in cutting gears and manufacturing transformers. One day John Keane — who'd worked with Jim Sr. on a summer job at NCR and was now working as a salesman for SREPCO[1] — stopped in Crown's offices to see Jim. Once televisions were in mass production, Keane told Jim, there would be a big market for antenna rotators. "What is an antenna rotator?" Jim asked. Keane explained that the rotator positioned the television antenna in the direction of the strongest television signal so the television had better reception. The Alliance Manufacturing Co. had been manufacturing such a rotator and had thousands of back orders for the product, said Keane. The following week, Jim got a rotator from SREPCO.

Using motors and gears from its heat regulator, Crown began to design a rotator of its own. Many nightly visits to Dayton followed as Bill Keuner, a retired Master Electric engineer, worked on the new design. With the Tenn-a-Rotor that Alliance was manufacturing, there was no way for someone inside the home to know which way the antenna on the roof pointed (and whether it was in the right position to pick up a particular signal). You could only estimate the location by looking at your watch, because the antenna revolved once a minute, the same as the watch's second hand. Crown felt that if it could find a way to indicate inside the home which way the antenna was pointing, it could produce a more acceptable product. So the firm developed a control unit that sat on the television and showed people which way their antenna was pointing.

Control unit

Growing sales of rotators reflected growing sales of televisions. In 1946, there were only 6,000 sets in the whole country; by 1949, one million. By 1955, when Channel Master made an offer Crown couldn't refuse, television was a national institution. As the fortunes of television rose, so did the fortunes of antenna rotators.

The tooling and technology to manufacture an antenna rotator were not unlike those used to manufacture thermostatic controls, so this was a natural transition for Crown to make. And clearly there was a potential market for the product. So in 1949 work began on the Tenn-A-Liner, a rotator with a directional reader. Crown still had service work coming in on its heat regulating equipment. They wrote to customers explaining that certain equipment was obsolete or no longer available. For the most part they were able to help people keep their old equipment working, but now they had to focus on their new product.

The first drawings for the original Tenn-A-Liner were dated June 3, 1949. The drive unit for the first model, the AD-1, was essentially a heat regulator on its side with a vertical shaft inside a housing. In other words, they took the standard damper motor from Crown's heat regulator, which had a quarter-inch shaft coming out of it, added a fitting with two tabs on it, put that in a sand-cast housing, and put a tube down the neck of the housing in which to set the antenna post. The thinking was, "We make our own motors, we've got gears, and it turns at about the right speed."

Remember, Crown was originally a sales company, not a manufacturing company; it took them a while to become sophisticated about manufacturing. Old-timers reminisce with amusement about those early days of development. As they were developing the rotator's control unit, for example, Wilbert Will pointed at the control and speaker housing on Carl Dicke's desk intercom and said, 'I like that shape. Let's make it like that.' Carl agreed, so the control unit on the first rotators ended up looking a lot like an intercom.

Crown's first foray into industrial design was with a rotator knob that had indentations for your fingers, a design they got the local high school art teacher to help them make out of clay. Everyone involved forgot to take into consideration that when the clay dried, the dimensions would change. Crown learned a lot through trial and error.

Women on line

From early on, men ran Crown's machines and did the heavier work, but women did most of the assembly work on antenna rotators. When Crown had 280 people on the payroll, two-thirds of them were women. By law, they were permitted to lift no more than 27 pounds.

Crown was making antenna rotators but had a problem with quality and, in the beginning, did not have anyone responsible for quality control. Forty people were doing everything: engineering, manufacturing, selling, and accounting. In later years, Jim Sr. would say, "It was harder to get from one employee to ten than it was from ten to a hundred," and the early employees knew exactly what he meant. They had started from scratch and didn't know exactly what they were doing. Everybody in the company was starting over, and everything they did was experimental. But to tell the truth, getting from ten to a hundred employees was not that easy, either.

The rotator's drive unit used a damper motor inside an aluminum box. The main flaw of the unit was that the motor — designed only to raise and lower a couple of small furnace doors — was now expected to turn an antenna on the roof. The motor was too light, the gears weren't heavy enough to withstand the torque of the wind, and on top of that, condensation tended to rust out the works in the unit on the roof. The rotators "made their own condensation," says Jim Sr. "They would be cool in the evenings, then the sun would come out and hit the unit, the unit would sweat, the sweat would run down into the inside, and we'd get a cupful of water out of it, and for a long time we couldn't figure out where it was coming from." To solve the problem, Crown developed a loose-fitting bottom cover to allow the condensation to drain out.

A fellow with an office on the top floor of a building in Lima, Ohio, had been calling to bad-mouth Crown's antenna rotator. Someone went over and brought his rotator back to Crown, where everyone could see that the quarter-inch shaft coming out of the old damper motor was twisted around like a drill. That's when they knew they had to develop a totally new rotator, a heavier model.

Working with Carl Dicke, Wilbert Will came up with a drive unit design that featured a new shape for the sand-cast aluminum housing, plus a mainplate with a motor, gear train, and a mast assembly on it. Engineering later designed a new die-cast aluminum mast and housing, but the inner mechanism for the rotator is still essentially the same one Wilbert Will designed then; only the sending unit is different.

Strategy

After the death of her husband, Carl, Irene Dicke used the $25,000 she got from life insurance to buy machinery that she leased to Crown. This gave her an income and helped Crown at a time when it was starved for capital. It was a wise investment: Despite a long period when she needed private nursing care, she died with a comfortable estate.

Although Crown's original directional indicators — made by the Triplett Electrical Company in Bluffton, Ohio — had good dial faces, they had been too expensive. No more than a thousand of those first units had been sold when Allen Dicke asked a friend at the King-Seeley Co. to design a different dial face. The indicating unit on the new rotator was based on a Ford gas gauge King-Seeley manufactured, adapted to indicate North, East, South, West, North (the directional counterparts of Empty to Full) as the antenna on the roof turned. The new unit cost Crown only $1.30, less than a quarter the cost of the Triplett indicator. And King-Seeley's daily production was so high that Crown's six-month supply went out the door in under twenty minutes.

The first control housing was sand cast aluminum[2] because the tooling for sand casting was inexpensive. Workers polished the sand castings with big polishing belts, then painted the housing brown with an orange-peel texture. Later, when Crown could afford a mold, it changed to a Bakelite housing. The Bakelite gave off a pleasant aroma that old-timers still associate with Crown's early days.

Developing a suitable housing and mast for the rotator was one of Crown's biggest design problems. Originally the mast was sand-cast aluminum, but the first casting was not flexible; when the wind blew, the mast often snapped. So Crown went to an aluminum magnesium alloy, almag. Improving the quality was a painful process, especially for people in the field. On a sales trip out west, a salesman named Moulton took along what he thought were samples of the alloy casting, which he handed out at a meeting of dealers. While he was talking, he heard a snap, and laughter, another snap, another, and another, and more laughter. Crown had given him the regular sand casting instead of the almag and he was embarrassed to death. "We've got to solve this problem," he said when he returned to New Bremen.

Crown eventually had to decide whether to go from sand

casting to die casting of the rotator drive unit. With sand casting, the cost of labor and casting was high, but the tool cost for pattern equipment was low. With die casting, labor and casting costs were low (closer tolerances reduced the need for machining) but the cost of tools was high and delivery of the die-cast molds was slow. Sand casting was called for when you had low to medium demand for parts; die casting was appropriate only when you got into high-volume production.

Growing pains

By the late forties, needing additional space for the work coming in from Dayton, Crown built a one-story cement block addition along the north side of Plant I, the old hardware store. Everybody pitched in and finished it in 35 days. The old employees say it would take that long today just to get the permits. Above, Lester Howe begins the project by digging the footer.

When you're manufacturing something, either you have to buy the machinery and tools to make the parts or you have to farm the work out and have it done elsewhere. You have to reach a certain point in production before you can justify purchasing expensive manufacturing equipment. A creative businessman, Jim figured out a solution that not only got Crown some of the machine tools it needed to be able to grow but also provided an income for his newly widowed mother. When Jim's father, Carl, had died in 1952, everything he owned was invested in the company; he had even borrowed against his house for funds. The only financial asset in his estate was a $25,000 life insurance policy that went to his wife, Irene. Jim found himself with a 50-percent partner (his mother) in a company that had almost no net worth, and, to add to the problem, he was also her sole source of support. Irene Dicke sold her house and moved into a rental property until she was financially able to build a retirement home, which she did in 1955. She stayed in that house for the rest of her life. On Jim's advice, she had used the $25,000 she received in life insurance to buy machinery that she leased to Crown. This provided her with income and helped the young, capital-starved company. In 1959 Irene sold her half of the company to Jim, but she continued to lease machine tools to the company. For many years, she was also Crown's corporate secretary. Jim had a flair for turning negative situations into positive ones.

After a few rough years, Crown got its rotator up to a suitable level of quality and began to see pretty good sales for a small company. It had gone into production cautiously, starting with 10 rotators a day, but had quickly upped that to 25 a day. Each increase in production was painful, as new assemblers had to be trained. And competing with Alliance for a share of the U.S. market was tough: Crown's minimal advertising budget was meaningless against the $7 million a year Alliance was spending on ads. Crown didn't have the production to support that kind of budget, but its sales had been strong in Canada from early on, so the firm decided to launch its marketing campaign through the back door, spending all of its advertising money in Canada. In those days, as you drove down any street in the United States you could count ten or twelve Alliance units for every Crown unit; in Canada, the ratio was reversed. Crown established a subsidiary in Kitchener, Ontario, and in 1963 built a plant in St. Thomas, Ontario, to assemble rotators from parts shipped from New Bremen. It continued making Crown rotators there until the late 1980s.

Between the antenna rotators and the subcontracting work Jim was taking on, Crown was busy enough to begin to feel pinched for space. The second floor of Plant 1 — originally a thirteen-room apartment, but now one large space — was all for assembly, first of thermostats and then of antenna rotators. The hardest part of young Harold Stammen's after-school job was carrying supplies from shipping and receiving to upstairs, past the women working there on antenna rotators, to the stockroom. "I was always embarrassed," Stammen recalls. "I had to pass all these women, and of course they all watched whatever you did."

In 1947-48, Crown built an addition along the north side of Plant 1 to house punch presses, screw machines, and an enlarged toolroom for Wilbert Will. Every man on the payroll was asked to pitch in and help build the one-story cement-block building, and they got it up in thirty-five days. Decades later, many of them still remember carrying cement blocks and speculate that nowadays it would take thirty-five days just to get permits. Crown needed the new addition for all the work coming in

Snow job

Long before Crown purchased the house of the Ekermeyer sisters, Crown employees knew it well. One day Maude Ekermeyer called Craig Hirschfeld to say, "I just paid a boy $1 to shovel my sidewalk to the curb. Now the state plow has come through and plowed the snow back on the sidewalk. I think Crown should come over here and clean it off." When Craig asked why Crown should clear off snow left by the state plow, she said the only reason the state cleaned Route 66 was for Crown's convenience. Amused by her logic, Craig sent someone over to shovel the sidewalk.

from Dayton. Once it was finished, says Wilbert, "they bought a milling machine that we used as a jig bore and we started making all the tooling for Crown." They set the addition back a couple of feet because Allen Dicke didn't want it to go all the way to the Ekermeyer property line. (Irene and Maude Ekermeyer were two women who were accustomed to peace and quiet. Every time there was a little noise, or a light bulb went out, or a window broke, they would call the plant.)

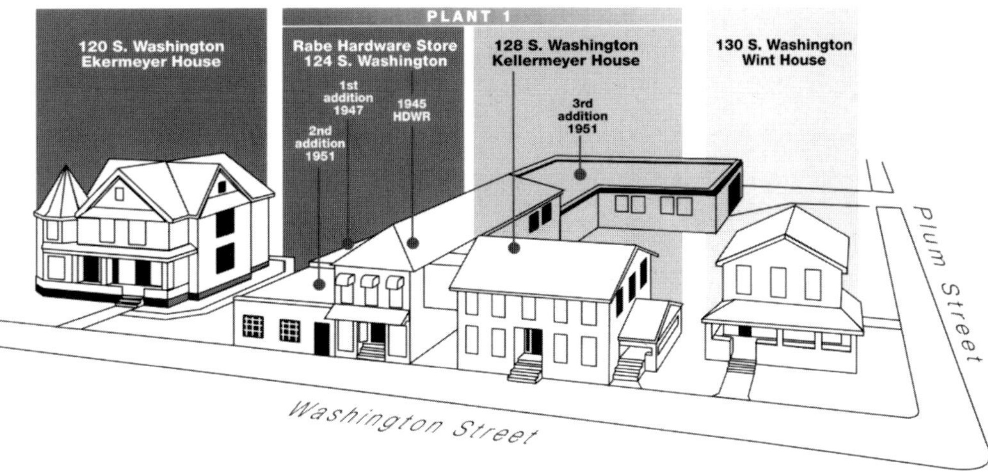

Adding on

In 1951, Crown extended the first addition on Plant 1 forward to the sidewalk and built a third addition in back of the plant, a cement-block building replacing Sophie Kellermeyer's beautiful garden, to handle shipping and receiving. The north (side) addition goes all the way to the Ekermeyer property line, reflecting a change of heart and management.

In 1951, they extended the north addition forward to the sidewalk. They also built another addition to Plant 1, to handle shipping and receiving: a cement-block building south of the hardware store and behind Sophie Kellermeyer's house, replacing her beautiful garden. Allen Dicke had left the business and had sold the Kellermeyer house to Crown (Mrs. Kellermeyer had the right to live out her life in it), so the south addition went all the way to the Ekermeyer property line; you can see a jog where the firm had a change of heart and management. After the Ekermeyer sisters died, Crown bought their house; later it bought the corner house south of the plant from a Mrs. Wint. If the company was to grow, it needed space for workers to work. Eventually it owned most of the houses on the block.

For a while, the advertising department used part of the Kellermeyer porch. Soon it outgrew that and moved into the basement with engineering. After the new toolroom was finished, the men built a little second-story walkway to connect the upstairs of the Kellermeyer house with the upstairs of the hardware store. At first, Crown occupied only the upstairs and porch of the Kellermeyer house. The engineers (Paul Dicke, Bob Marshall, and Jim Uetrecht) were crammed into the porch at the end of the Kellermeyer kitchen.

By 1953, Crown had a whopping 68 employees and was bursting at the seams. Around that time, Jim ran into Verlin Hirschfeld, a fellow four years his senior with whom he'd gone to school. Jim and Verlin had gotten out of the military the same year and Verlin was working at the Arcade, a small department store a block north of Plant 1. Downstairs in the Arcade were four departments: groceries, men's clothing, appliances, and dry goods; upstairs was furniture. Verlin and a friend had remodeled the building. The department store "had a little bit of everything but not much of anything," says Verlin. The Arcade was having trouble competing with the shopping malls that were beginning to dot the countryside. Although the one closest to New Bremen was thirty miles away, everybody was already starting to go there to shop. Verlin and his partners decided to close their business while they were ahead financially.

"If business keeps going like it's going," said Jim, "we're going to outgrow our building. Do you want to sell this one?" Verlin had been authorized to sell and said he'd be glad to do so for $45,000. Jim didn't have that kind of money, but the two men made a deal: Jim would pay $5,000 down and $5,000 a year, at 4 percent interest, and Crown would have a second plant. Although Verlin's experience was strictly in retail, Jim asked him to come work at Crown. Verlin said he knew nothing about that kind of business, and Jim assured him, "You'll learn. We'll teach you whatever it takes." So Verlin, who had a wife and two kids, took the position for $1.40 an hour and spent a couple of months in the plant to learn what went on there. Once in a while Jim would come into the plant and say, "Hey, Verlin, at ten o'clock I've got this insurance guy coming in. Why don't you run home and change clothes?" Verlin would do so, changing out of coveralls into a business suit, and join Jim and whoever he had a business meeting with that day. Slowly he learned what he needed to know.

By 1956, Canada was a significant part of Crown's market and Crown was sailing along, making about 85 rotators a day. One evening, Tom Shelby, Crown's sales manager, ran into a fellow named Jack Launer in the bar car of a train. The fact that Shelby was not much of a sleeper was to improve Crown's fortunes. As the two men chatted over drinks, Shelby mentioned that his firm made antenna rotators. As it happened, Launer was a buyer for the Channel Master Corp., an Ellenville, New York firm that was the world's largest manufacturer of television antennas. "We used to buy rotators from some guy," Launer told Shelby, "and we got a really bum deal and a black eye." A unit Channel Master had bought from another rotator manufacturer had failed miserably in the field, costing Channel Master a bundle, and they were looking for a new and tested design. "We could probably talk business," Shelby told Launer. So in a bar car on a Pennsylvania Railroad train someplace between Chicago and New York, Crown took a first step toward its first major deal, one that would ultimately take the firm in a totally new direction.

Channel Master gave Crown an order for 15,000 antenna rotators — at a time when Crown was making only 85 rotators on a good day. This was beyond belief. And shortly after Channel Master took the Crown rotators to the Chicago Radio Parts show, the order went up to 25,000. This sent the firm into a panic. How could Crown step up production from 85 rotators a day to 500? They did it, but first Crown had to gear up to meet the vastly larger order, reorganizing production, rearranging and adding assembly tables, rerouting the conveyor system, training new workers, and putting someone in charge to make sure everything moved smoothly. Al Schwieterman, production manager at the time, said, "There is no way I can do this without good parts."

Quality control was still a problem at Crown. To improve the flow of production and make smoother and quicker assembly possible, manufacturing needed to produce standardized, interchangeable parts. Someone suggested that the purchasing department should buy things, and the machine shop and punch press department should make things, only according to drawings. This was before Leo Schulte was around, or Leo would have said, "Hey, if it was all according to the drawing it definitely wouldn't go together." To the idea of using only parts that measured the same as the drawings, Carl had responded, "You'd shut us down right now." Crown had a long way to go before its drawings would be suitable for precision manufacturing.

Looking good

The money they made selling the Hershey Motor Stoker business allowed Jim and Eilleen to build a new house, which impressed the Channel Master people enough that they upped their distribution offer. Crown's production numbers doubled, then tripled—and forty years later rotators were still going strong.

Manufacturing standardized parts also requires adequate inspection. The need for better inspection had become evident earlier, when Carl Dicke was still alive. Hammers, pliers, and other objects that should not have been on the assembly line could be found there (you can't have hammer-free assembly when the parts aren't precision-manufactured). Carl had resisted the idea, but on Jim's authority Crown bought calipers, a pair of micrometers, and a few other measurement tools and enclosed them with chicken wire in some idle space beneath one of the conveyors, after adding a door and building in some shelves. That was how Crown's inspection department got started. Then, after the cement-block addition was built, serious inspection began, with Annabelle Wagner as the first inspector followed eventually by chief inspector Norman Ruley. Soon Crown had not only strengthened inspection but had begun making all of the rotator parts itself (except for screws, steel, and plastic parts) so it could control cost, quality, and reliability of supply. Jim Sr. wanted Crown to be as self-sufficient as possible, as soon as it could afford to be.

After Channel Master had placed its big initial order, representatives from the firm came to New Bremen to negotiate an agreement with Jim Sr. Here luck and timing played a part in Crown's fortunes. In early 1957 Jim and Eilleen Dicke lived in a modest house at 107 North Franklin Street. In July they moved into a fancier house they had built on East South Street with the money they made from selling the Hershey Motor Stoker business to Blue Coal. Jim II, who was finishing the sixth grade at the time, remembers his father saying that if they had still been in the other house they wouldn't have been able to make as good a deal: "The Channel Master people came to New Bremen thinking there wasn't going to be any problem making a deal with this little manufacturer, and when they saw that Mother and Dad were living in a really nice new house they thought, 'Oh gee, these people are better off than we thought.' So they offered him a better deal."

Essentially Channel Master wanted to take over the marketing of Crown rotators worldwide, except in Canada. The advantage for Crown was that Channel Master had distribution and sales outlets that Crown could not easily duplicate. Crown couldn't hope to have the same kind of market muscle as Channel Master because Channel Master was selling a full line of products, of which the antenna rotator was just one. After making the deal with Channel Master — there was never a contract, only a handshake and good faith — Crown's production numbers doubled, then tripled.[3]

In 1954 Crown had moved its general offices into the Arcade, now called Plant 2, which for a considerable while retained the look of an old small-town department store — with a space heater in the corner of what is now the lobby. What is currently Crown's Plant 2 office building at that time housed everything from engineering, sales, and executive offices to manufacturing operations. Engineering took the basement, mainly because it was the only place in the building that was cool. That it was already full of shelves from the department store was also convenient. On the main floor, the first room to the south (formerly groceries) became office space; the second room (men's clothing) housed sub-assemblies and the parts being made for the rotator division. The third room (formerly appliances) was used to make foul detectors for bowling alleys (a contract job). The fourth and northernmost room (formerly dry goods) now housed automatic screw machines. These screw machines made a terrible mess; a decent room was getting splattered with oil. To counter the mess, sheets of plywood were laid on the lovely hardwood floors. A men's restroom was built inside the old Arcade meat cooler, where it remains today.

In 1959 and 1960, Crown's first walkie-stacker lift trucks would be made in the north end of the building, where the screw machines had been. Lift truck assembly and painting would also be done in Plant 2. Upstairs (where furniture had been — where it was unheated in the winter and uncooled in the summer) Jack Lawler would soon manage Repair Services. "When I first came here in 1965," says Carol Jones, associate vice president, "the office floors were still covered in green linoleum, even in Jim Dicke's and Tom Bidwell's offices." The linoleum gently rolled up and down from section to section to bridge different floor heights, and manufacturing still cohabited with the executive offices. The building was not taken over completely by offices until the late 1970s, when the north end of the building upstairs was converted from storeroom to offices for Crown's international department, of which Jim II was then vice president.

When transistor radios came out in the late 1950s and

early 1960s, they were selling for eighty dollars, which was expensive at the time. Still, Channel Master did well with them, especially in Canadian provinces such as Saskatchewan, which had long cold winters and often no electricity. Battery-powered and portable, the transistor radio was immediately popular with farmers and then, as the price came down, with everyone. Channel Master was importing its radios from Japan but needed to set up a service system. They asked Crown if it would repair radios for Canada on a contract basis. Bob Nieberding flew to Ellenville, New York, with Eddie Ritz, the plant manager from Crown's St. Thomas, Ontario, plant. Eddie had to learn about transistor radios, see what kind of repair facility existed in Ellenville, then duplicate it in St. Thomas. By the 1960s, a fair proportion of Crown's revenues would come from such contract work, for both the private sector and the government.

Crown's volume jumped with Channel Master's marketing capability, but after a little more than a decade Crown began thinking that the antenna rotator market might die out just as the coal-based heat regulator market had. Crown engaged a market study from the management consultants Booz·Allen & Hamilton, and in a 1968 report Booz·Allen confirmed Crown's worst fears: The growth of satellite communications and cable television would probably lead to the demise of the antenna rotator, so Crown should diversify. Booz·Allen was right about the growth of satellite and cable communications, and Crown did diversify, but luckily it never stopped manufacturing the antenna rotators. Nearly thirty years later, rotator sales remain strong, the Crown rotator is still one of the best on the market, and Plant 1, the former hardware store, is still producing rotators — in pretty much the same manner it did then.[4] Everyone is surprised that sales have remained as stable as they have. Many rotators are shipped for sale outside of the United States, but a substantial proportion of sales is still in this country. Although the satellite dishes popular outside of urban areas pick up hundreds of channels for less money than it would cost to pick up only twenty channels on cable, they don't pick up local programming, so for good reception of local channels a household still needs an antenna. And people who have cable often use the rotators to improve FM, CB, and ham radio reception.

Crown's alliance with Channel Master quickly tripled Crown's sales volume. By the end of 1957, however, Crown had a product to manufacture but no product to market, and Crown was historically a marketing-oriented company. For a while, much of Crown's sales energy would go into digging up contract work. Contract work would help Crown meet payroll at the same time that it helped prepare the firm for a shift into a totally different market. ❍

[1] *Standard Radio and Electronic Parts Company, of Dayton, Ohio.*

[2] *This aluminum housing was made in Sidney, Ohio, by Ross Pattern, Foundry and Development Co.*

[3] *Crown retained the patent on its rotator, continued selling its Airline unit to Montgomery Ward ("Monkey Ward" as they called it), and sold the Crown unit all over Canada. Eventually Alliance, their chief competitor, got out of the rotator business.*

[4] *Except that, to prevent carpal tunnel syndrome (a problem that often results from repetitive motion), the employees rotate jobs regularly. One result of this rotation is that everybody has multiple skills and nobody does the same thing all day.*

Carl and Jim, fishing *As a young boater (facing page), Jim had earned the nickname "the brat of Lake Loramie." Parents of rambunctious boys, take comfort: When the time came to shoulder responsibilities, Jim was up to the task—and his sense of adventure proved useful in business.*

The Man Who Made Things Happen

(1945–51)

4.

There's something almost contradictory about the things employees say about Jim Dicke Sr., and yet when you meet him you know what they mean:

"He was very impressive — just his presence. You knew he was the boss."

"He was a solid down-to-earth person."

"He was willing to ask a dumb question."

Desk time *A well-rounded businessman, Jim knew that sales, finance, engineering, production, and purchasing must all work toward a common goal or the company would not succeed.*

"Jim didn't let a lot of grass grow under his feet."

"He's a good listener."

"He's gifted at reading people."

"A strong company head."

An outgoing person with a good sense of humor, Jim Dicke knew everybody who worked at the company in those first decades, and he always had something to say to them. As he walked through the plant, he would ask how their families were or how things were going and he always seemed to know in detail what they were doing. "That always impressed me," says one employee, "how he knew exactly what I was working on and how I was progressing. I sometimes swore he must have come in at night and gone through my desk to figure all this out. I could never understand how he stayed so in tune. He had that kind of awareness throughout the company." He was detail-oriented up through the mid-eighties: He liked to know what a part cost, why it was being scrapped, and how X affected Y and Z.

That personal touch extended to every department. "He would stop in the offices," says Mike Spicer of advertising, "and talk to everybody. He would go over to your secretary, pick up some pencils on her desk, and sharpen them for her. The whole family is easy to work with. They want you to give them the ideas. They don't usually give you an idea to start with."

"And he wasn't just walking through the factory being friendly," says another Crown employee. "As Jim walked through the plant greeting employees, he would often ask them how things were going. An employee might say to Jim, 'Things are fine but this fan keeps blowing on me. It really creates a draft.' Jim would ask a foreman or someone else, 'Can you do something about that fan?' In an hour or so there would be a fix. Workers really appreciate that. He encouraged speedy, credible responses, when appropriate, and when Jim wanted something done, it got done. He would let you know it should be done, although not how, and then he would leave you alone and come back the next day, and by golly, you had it done."

People responded to Jim's requests when they weren't even expressed as requests — when they were mere hints. Once some contractors were extending a mezzanine to add another work bay. There was a furnace at the end of the existing bay so they decided to put another wall just past the furnace. Jim walked through and noticed that the oak floor they were putting down didn't go up to the oak floor in the other bay. "It sure would have been nice if we could have run that floor all the way through to join with the other flooring," he said. They had already started putting up the partition, but they immediately tore down the wall and extended the floor all the way through.

Jim Sr. was a well-rounded businessman who understood that all elements of the business — sales, finance, engineering, production, and purchasing — must work toward a common goal or the company will not succeed. He didn't have to function as a committee with a committee's compromises and watered-down decisions and revisions. He was a strong company head.

He was strong even while his father Carl was still alive. Jim became vice president after he and his father bought Allen's share of the company, but as president Carl still had firm ideas about how things should be done. Carl traveled back and forth between New Bremen and the company's hunting lodge at Medix

Run, Pennsylvania, where he spent a lot of time. "Carl would tell me to do something that Jim didn't want me to do," recalls Paul Dicke. "I would say, 'Geez, when Carl comes back I'm going to catch hell for this, you know,' and Jim would say, 'Let me handle that when it happens.' And sure enough Carl would come back and whatever it was, he would just blow his top. After a while he would take off again and we would keep right on going. He threw up his hands and said, 'I guess my son is going to run it the way he wants, even though it's wrong.' Of course, it wasn't wrong."

"The gentleman is marvelous," says Arnie Heitkamp. "He had an uncanny ability for picking good people. He's gifted at reading people. In my opinion you are born with that quality or you don't have it. That's why some businesses are successful and some are not."

Jim was a hands-on person, interested in all aspects of the operation and what everybody was doing. He never forgot anything and was interested in everything. In that sense, he was a good role model for the managers who came up under him, who, even even after Jim's own personal inspection tours tapered off, continued to make sure the restrooms were being maintained and the plant being kept clean.

"Jim had a way of putting you at ease," says Harold Stammen. "He could be open talking to you and you felt open talking to him. I think all of that is part of Crown's success. People felt listened to with Jim. If he didn't agree with something you said, he was forthright and gave the other view — why Crown doesn't do this or can't do that. He was a good listener, he picked things up quickly, and he was very quick at responding with a good, credible response. He was a very fast thinker, which always impressed us. Sometimes I wished he would go a little slower. I'm a slow thinker and he was a mile ahead of me. He also moved physically at a very fast pace. When he came through and said 'Hi' to you, you sometimes didn't have time enough to think about how to respond. He was on the move all the time. But I also know that if you sat down with him he was patient because I was in meetings where he was present and he was a very good listener."

Number one

Jim Dicke loved running Crown. "Always try to hire the best people you can," Jim Dicke said. "Hire somebody you think is even better than you are because the best way to get ahead is to have people working for you who are so smart they're going to make you look good."

He had a gift for getting good people to do good things and feel good about it. He wouldn't ever say, "Do this." He would say, "I'd kind of like this," and people would figure out how to make it happen. He realized that you can get somebody to work a lot better if you don't tell them what to do or how to do it, if you simply say, "This is what I like." They feel more a part of it then, the way children are more motivated to do something when their intent is to surprise you than they are when you order them — not only to do it, but to do it right now, and your way. Jim had a way of couching things so that you felt you were going to surprise him with the result. He knew how to motivate people, knew how to "empower" them before that word came into vogue.

He also had a knack for converting the local can-do spirit into teamwork; he was a team-builder par excellence. In later years, when he had a management team, he always encouraged them never to be afraid to hire somebody smarter than they were. "The best way to move up the ladder in the company," he told them, "is to train your successor. If you have people who can take over your job, then you can take on more responsibilities and keep moving up. If you want to continue to sit there and do your job and not teach anybody, or not let anybody learn from you, or not have anybody backing you up, that's where you're going to stay."

Jim believed that a company director or owner's most important quality was to be willing to listen to other people, to be willing to accept suggestions, and to be willing to give credit. "In a business discussion," he told them, "if you make a statement and somebody else brings a new thought to the table, you should, one, be ready to accept it and, two, be ready to give them credit for having brought it up. Don't say, 'I don't think we should do that' and two days later have them find out you're doing exactly what they suggested under your handle and your name. That's a good way to kill support and enthusiasm. Most people would say that at Crown we're willing to change our minds, as top management, and give credit to who came up with the suggestion. Many times it's a combination: part of an idea comes from here and part of it comes from there. Then you always have to say 'we': 'We thought this out,' 'We're going to do this,' or 'We came up with this idea.' So many people kill incentives by using the word 'I.'"

Jim created this feeling of "we" from top to bottom. "One thing that helped keep productivity up at Crown in the early days," said Verlin Hirschfeld, "was that every day that Jim was in town he made sure to spend a little time in the plant, where the action was. Jim was pretty mechanically inclined without being afraid of who was making what, and if he was talking with an operator about why something was being done a certain way, he'd say, 'Maybe we ought to make this thing, or go a little slower or a little faster, move it over a little bit.' He had a knack for talking with people about how things worked. Together they might come up with a better idea of how to do it. As he went through the plant, everybody knew him — 'Hi Jim' — and he knew them, by first name. They weren't numbers; they were people. Somehow that meant something more to people than that extra ten cents an hour." ❍

The clean room *Paul Dicke (left) works with Bob Marshall in the dust-free, temperature-controlled "clean room" required to manufacture precision parts for the military. What possessed Crown to take on contract work it had never done before? A can-do spirit—and the fact that if they didn't work on something, they wouldn't get paid. On facing page, a Crown worker winds coils.*

Learning by Doing: Subcontracting *to the* Military

(1950–74)

5.

In the 1950s, as Crown was backing out of the heat regulator market, launching its line of antenna rotators, and looking earnestly for new products to develop, its survival and growth were greatly facilitated by a period of intensive contract work, manufacturing and repairing mechanical and electronic components for private industry and the government, especially the military. Over two decades and more, Crown manufactured and repaired hundreds of different products for the Army and Air Force, first as a second-source subcontractor and then on a direct basis. At the time, some people at Crown viewed much of Crown's contract work, from the late 'fifties on, as wasting time when it should have been developing the lift truck, a promising product that it launched in 1956. In fact, the contracting work together with the antenna rotator line subsidized the lift truck's development. Even a brief period of producing novelty items (between 1954 and 1957) helped Crown develop into a manufacturing firm capable of producing first-class parts and products.

Much of Crown's contract work reflected a period of heavy defense spending by the U.S. government. In August 1949, the Soviet Union had tested an atomic bomb, and in 1950 the United States, having adopted a strategy of containing Communism within Asia's mainland, had sent troops to Korea in a "police action" that would later be known as the Korean War. There was an active Air Force base in nearby Dayton, and as the Cold War between the Soviet bloc and the West heated up, Crown found itself manufacturing many products and fulfilling other small contracts in support of the U.S. military. Some of Crown's subcontracting work was initially brought in by Fred Stork, an industrious sales representative who had gone to school with Jim Dicke and who didn't know how to take "no" for an answer. Stork helped Crown find work with manufacturers who did business with Gentile Depot, a government procurement agency near Dayton, and those manufacturers took Crown's work and packaged it in special government-specification long-term storage materials. After Crown got established as a subcontractor, it began working directly with government buyers.

Carl Dicke was still alive when Crown took on its first subcontracting job, to manufacture a radar phase shifter for the Hoffman Manufacturing Company of California. The phase shifter was a device the size of a yo-yo that was part of the radar equipment that went into the nose of an airplane. The phase of a radar signal was shifted 180 degrees as it passed through the phase shifter. Mind you, almost no one at Crown knew how the phase shifter actually worked, or why the phase had to be shifted, but they didn't have to know how it worked to make the part; they just had to know how to read the blueprint, make the part, test it, figure out if it passed the test, and, if it did, ship it.

In 1951, a few months after leaving his job at Ford's garage in Sidney, Ohio, a young engineer named Bob Marshall joined Crown's engineering department. Jim told him they were having trouble meeting the delivery schedule for the phase shifter. Getting things moving was right up Marshall's alley, and he was soon organizing for more efficient production. Marshall told the fellow running the machine, "Shut that thing off; I want you to inventory every part you've got here." From the inventory, Marshall figured out the ten parts needed to start making deliveries and had Purchasing order them. "The problem was, nobody knew what we had, whether it was good or bad, or what we needed. It was just sort of haphazard." The idea was to get everything in line so you knew what you had and didn't have, a principle anyone who has ever cooked a meal understands. "At the time, they were relying on one person to do everything. That person was the machinist, the assembler, the jack of all trades — and he was a little confused." Marshall applied a new approach called systems analysis and got things rolling.

"It was tough," says Bob. "Crown had a jerry-built system, including a lot of machinery that used to belong to somebody else and had already worn out. One fellow at Crown probably kept us in business: Wilbert Will. I don't care what you asked Wilbert, if it was anywhere reasonable, he did it. If you could get two things to go together and work, he'd do it. He was not a trained engineer, but he had the moxie to get the job done. Wilbert's brain seemed to go a mile a minute. He would be listening to you trying to explain what you wanted — you wouldn't even necessarily know what you wanted but you would tell him and he could pick up on what you were trying to say — and an hour later he had the job done."

Production was occasionally makeshift, the way it almost has to be when a company is growing faster than the space to accommodate it. During the summer of 1954, two young employees worked in the crawl space beneath Plant 1, a cool workspace for welding the phase shifter. Earlier, they had gone under there with a small air hammer and a pick, carrying the dirt in buckets up a short stairway. A portion of the crawl space was made deep enough that they could go in in a squatting position; they sat under the floor on low chairs. "I don't know if our heads were between the floor joists or what, but there was no more than a five-foot clearance," says Harold Stammen, who was then a mere teen-ager. "But it was cool in the summer. We generated a lot of heat because we used a torch, and there was the flux and then the silver solder itself. Two people worked together on the project, one person soldering and the other doing the cleaning and prep for the next part."

In 1953, Crown got an important second-source contract:

Maestro of the tool room

Whenever you wanted something special done, you went to Wilbert Will. Wilbert was one of Crown's mechanical geniuses who never went to college. He was in charge of the tool room almost from Day One. "It was right down my alley when Crown put me in the tool room. I was always tinkering with something at home. I don't know of anybody in the Crown tool room who has been to college." Anybody with a problem would give it to him and, says Wilbert, "I generally came up with an answer. Very few problems I couldn't solve." In retirement, he began building an airplane. He still does much of the restoration work on antiques and decorative items for Crown's guest houses.

to build turn-and-bank indicators for Navy aircraft. Fearful of atomic attack and mindful of the need to diversify its manufacturing base, the government had established small-business set-asides to make military work available to small manufacturers in rural areas, so that destruction of a key manufacturing area wouldn't wipe out all of the military's industrial support. Crown got its Navy contract through such a set-aside.

What was a turn-and-bank indicator? Visualize the face of the speedometer on an older-model car, with a dial that tells you how fast you're going. A similar hand on the black face of the turn-and-bank instrument moved to tell pilots they were turning so many degrees per minute. If they banked (tipped) the airplane to the right without enough rudder, a little ball in a curved tube would roll to the right to show that the bank angle was inappropriate for the turning speed. (The indicator looked like a smiling face.) With proper turning and banking, the little ball stayed in the center.

To produce the turn-and-bank indicator, Crown had to improve its manufacturing facilities. For one thing, it needed Excello machines, special state-of-the-art precision boring machines produced by the Excello Manufacturing Company. Excello machines were crucial for military work and were "on allocation" because they were in short supply. The government had promised Crown two Excello machines to manufacture the turn-and-bank instrument, but General Motors had had a fire in its transmission plant. All of their Excello machines were damaged, and GM wasn't about to let Crown get in line for the machines in front of GM. "Jim went to Washington," says Bob Marshall, "and when he got back we had our Excello machines. For a little company like we were to buck General Motors, to tell the government, 'Look you can't do that to us. Two machines aren't going to make or break General Motors but they are going to break us,' that's what Jim was good at. I don't know what the consequences would have been if we hadn't got them. General Motors was putting pressure on the government to get machines for themselves that we needed, but *we* got them." They installed them, and a machinist named Harry Boltz (a class A lathe operator) ran them, in a machining center established near the northern entrance to Plant 1.

Crown also had to develop a "clean room," a temperature-controlled, dust-free room. Paul Dicke visited the Davenport, Iowa plant of Bendix, the prime contractor on the turn-and-bank indicator, and returned with a story his colleagues shook their heads over: "Those people work inside a dust-controlled room. They wear boots over their shoes and a smock and some of them wear snoods to keep their hair from getting into parts." The people at Crown were skeptical about Paul's report, so he took Harry Gilbert to Iowa to show him what was needed. Then Crown set up a pressurized clean room that occupied the southern portion of the second floor of the Kellermeyer house. "It wasn't as good as the one they had out in Davenport," says Paul, "but it served the purpose."

Crown's clean room wasn't clean enough, however. Crown's turn-and-bank instrument "looked good," says Bob Marshall, but it just wasn't up to speed; it wouldn't perform the way the government wanted it to. So "Paul Dicke and I went out to Davenport again. The fellow in charge looked at the instrument we'd made and said, 'It's beautiful; you did a good job on it. There's only one thing wrong: It's dirty.' Shoot, we wore boots on our shoes, and smocks, and something on our head; we tried everything we knew to get them as clean as we could and then cleaner, and he says, 'It's dirty.' We went in their clean room and he took this thing apart and showed us that the critical parts had burrs on them from the machine process, and lint from when we would wipe them off, but you couldn't see this without a microscope. So we went back home and bought a microscope, and everything that was critical in that unit went underneath that microscope, and that ended our problem. It was that simple. You need a special cloth to clean parts and you have to have a special machine that washes parts and dries them (sort of like a dishwasher, to eliminate the moisture). Everything that is on there is supposed to come off the parts except the burrs that are put on there in machining; those you just have to put up with."

Over several years, Crown built and shipped perhaps a thousand turn-and-bank instruments to the Air Force Supply Depot near Dayton, as well as some of the components (such as gyros) that went into them. The turn-and-bank indicator brought in a lot of revenue and put Crown in a position to do precision manufacturing on other products. Production of the turn-and-bank indicator allowed Crown to lease turret lathes and beautiful grinding equipment and other brand-new heavy machinery that income from the antenna rotator alone would not have supported. Having this heavy metal-cutting equipment allowed Crown to manufacture parts to near-perfect tolerance. It also produced some true machinists at Crown. When the turn-and-bank contract was completed, Crown felt it was ready for anything.

Later, when Westinghouse's Lima, Ohio plant was on strike, Westinghouse was looking for an outside contractor to make gyros for Minneapolis Honeywell. Crown paid a call. Laying out the blueprints, the Honeywell people explained that they needed precise work with very close tolerances. "We'll have no problem with that," said the Crown people. "We have an Excello boring machine that bores perfect holes from opposite directions."

"Where are you guys from?" asked one of the men from Lima (a sizable town about thirty miles from New Bremen). "I've been through New Bremen a bunch of times and I've never seen anything like that." Crown explained that it operated out of a factory that looked like a small hardware store in front. Westinghouse became a regular customer.

A lot of Crown's contract work was government work, most of which it got through competitive bidding. The government would send out what it then called IFBs (invitations for bid), with a fixed closing date. You had to have your bid in by a certain date and the low bidder got the contract (usually a fixed-price contract). Maybe five or six companies would compete, so Crown would review the drawings, estimate the total cost (including the material cost to make each part, the time it would take to put it together, and the cost of packaging it the way the government wanted it packaged), come up with a bid that included overhead and profit, and hope it hadn't made any errors — so that when they got done they *would* have a profit. Having the right tools to do a certain job made it easier to come up with the lowest bid. Crown may also have had lower overhead and labor costs than other firms; Crown's average wage in those early days was about $2 an hour.

One of the most profitable and complicated products for

which Crown won a contract in the 1950s was for an antenna beacon, a brass component the size of an index finger, which would be attached to the outside of a sonobuoy. Sonobuoys were electronic devices used to detect submarines. Three or more of them would be dropped around an area where a submarine was thought to be lurking. The signals they transmitted would bounce off the sub, and the planes overhead would know where to drop depth charges.

At first they had a problem with the antenna beacons, says Jim. "They could not get them to work until they shotblasted them with glass pellets to get rid of small metal burrs left inside from machining and assembly." Women did most of the work on the assembly line. One woman drilled holes, a second woman inserted screws, and a third soldered the heads onto the antenna beacons. Crown built antenna beacons for about ten years, manufacturing hundreds of thousands for Magnavox, for Hazeltine Electronics, for E Systems, and for several other companies.

Crown did a lot of toolroom engineering work for Magnavox because Magnavox made nothing itself and did no machining. All it had was an assembly plant. "They bought everything on the outside but they were first-class price buyers," says Arnie. "They knew where to go, and if they needed something they knew how to get it." One of the first jobs Crown did for Magnavox was to make a braking device — essentially, a round beryllium-copper ring with a part brazed on the end — needed for a resolver or a syncro in "some electronic gizmo Magnavox made." The brake would stop whatever it was used on by dropping a pin on it to catch a raised portion of the ring. "I know you guys can make this thing," said Walt Rysick, one of Magnavox's top people, "and we gotta have them."

"It was a real weird deal," says Arnie. "Every time the government purchased them they paid a big price for them. But we didn't even make them at the factory. We would give Wilbert Will, our toolroom foreman, an order to make them at home and we'd pay him so much apiece."

Crown also manufactured the dial drive for a ship-to-shore radio system Magnavox was assembling for the government. "We had a hell of a lot of trouble with that," say the men who worked on it. It was a challenge because they had to make spring-loaded anti-backlash gears for the thing, and again they had trouble figuring out how to eliminate burrs from the complex gearing. Crown also made all of the brass tuner cases, radio-frequency cases, and oscillator buffer cases used in the military VRC-12 radio sets (later selling the same components to P.R. Mallory in Indianapolis).

Magnavox helped Crown win the contract on the tuner cases. Walt Rysick hoped Crown would get the job because he knew the quality of their work, but he saw that Crown's price was too high. So Crown's sales manager, Ernie Jameson, and Arnie Heitkamp went to Magnavox's offices in Fort Wayne, Indiana, where Rysick walked them step by step through Crown's manufacturing process and the time required for each sequence of the operation. "It doesn't take that" he would say. "Here's what it takes..." or "Now look at this..." Educated by Magnavox, Crown repriced its bid and got the contract. For several years, Crown also made hub couplers for Magnavox. This machined precision casting was used in a driver system for a gear box assembly.

Another of many electromechanical assemblies Crown worked on during this period was a motor and gear-train assembly for radar antenna movement, which required machining a Gimbal Wave Guide. "That was a hell of a beast," said one of the fellows who worked on the project. On a product like the Gimbal Wave Guide, it was essential to get parts to stabilize, to remove residual stresses in the metal. The Gimbal was a fragile magnesium or aluminum casting on which holes were aligned two different ways, and if you measured it twice it would measure differently the second time. The engineers discovered that if they stored these metals at 10 degrees below zero for three days, the metal would shrink, nullifying all internal stresses, so that when you brought them up to room temperature you could drill holes in them and the measurements would never change. After Magnavox lost the contract with the government, Crown sold the same components to Sperry Gyroscope in New York.

In a good year, there was a balance between projects that were troublesome and projects that were both easy and occasionally a "big deal." One of Crown's earliest orders from the government was for a pair of test leads. "That was a $28,000 order," says Arnie Heitkamp. "Man, that was big numbers back then, and it was only a lead wire with a terminal on each end. I was tickled pink."

Sometimes the contract work took on a comical aspect. Off in a room south of where Crown's main offices are today, for example, Crown once made foam rubber antenna boots for a military jeep. The firm had bid on producing an antenna lead-in, which involved machining parts that were then brazed and welded together. But to get that contract they had to agree to make an antenna boot, a long thing shaped like an elephant's trunk. Crown knew nothing about making foam rubber, but bid anyway. The materials supplier explained: "You mix ten parts of this, three parts of that, two parts of this, you pour this mixture into a mold and after several hours it will set. Then you remove the finished part from the mold." With that vastly oversimplified description, Crown worked out a price and got the contract. Only later did they learn that the molds were three feet long and the process was an enormous hassle. Two men flew to New York to get first-hand information on how to make the thing, then Crown set up shop and entered a period of experimentation that sometimes resembled playtime in kindergarten. Using master shipping cartons from Wheaties or other large cereal boxes, they built a "heating oven" in which 100-watt light bulbs were used to speed up the cure time. They made five or six two-part aluminum sand-cast molds into which to pour the compounds. They would pour a portion of the mixed liquid into the mold bottom, C-clamp the top half of the mold onto the bottom, put the molds in the cardboard oven, and give them an hour or two to "set." After they removed the contents of the mold, they would coat the inside of the mold with Johnson's paste wax (which acted as a release agent) and get it ready for the next "boot." It was a messy job, but after eight weeks of experimenting they finally got their first satisfactory product. At that point Verlin came into the office and said, "Hey, we've got to quit playing around back there. We've got to start building units and ship this stuff." A couple of weeks later they got production rolling and met their shipping projections.

Sometimes the contract work was deadly serious. Until

1994, for example, the Strategic Air Command had B-52 bombers in the air somewhere at all times, as a precaution against nuclear attack. At one point, however, something began causing problems with the bomb hoist in the SAC bombers. The culprit appeared to be the brush holder, which held the brushes of an electrical circuit against the commutator of the motor that operated the bomb hoist elevator. Crown was led to believe that the electric motor on the bomb hoist elevator could catch fire; this had happened at least once when there was an atomic bomb on the elevator. (Crown learned years later that apparently all atomic bombs on the B-52 could be shorted out electrically.) Briefly, all the SAC bombers in the U.S. military were grounded because of this brush holder; the military had to tear out all the brush ring assemblies and replace them.

After Crown won the job on a competitive bid, the Air Force said, "Build them as fast as you can build them. We'll pay you a premium for every day you beat the delivery schedule." The product was needed so urgently that the supplies Crown needed became top priority, which meant that raw material suppliers were not permitted to ship to anybody else until Crown had its material. Generals and colonels were on-site at all times to make sure Crown was moving ahead with production. The toolroom worked around the clock to build the dies to make the stampings (round irregular-shaped parts and assembly fixtures), put them together, and get them to the government. Eilleen Dicke personally made sure there was a constant supply of sandwiches and coffee. Crown made perhaps 25,000 of the brush holders altogether, so the military would have extras to keep in stock. The employees worked 24 hours a day, seven days a week, to produce a supply of new, improved brush holders, and the B-52s started flying again.

First lady

Employees worked 24 hours a day, seven days a week, making parts for the Air Force's B-52s. Eilleen Dicke kept the coffee and sandwiches coming.

Another urgent Crown contract was to make heavy-duty shock mounts for underground phone cabling systems for the Bell Labs in Whippany, New Jersey. The idea was that in case of nuclear attack the shock mount would cushion the blow to the electronic equipment mounted on it so that it wouldn't knock out the country's communications system. The shock mounts Crown made were installed in Alaska — part of the Distant Early Warning radar line (or "DEW line") stretching across Canada to track incoming enemy aircraft — and probably all over the East Coast. Everybody at Crown who worked on this project had to have security clearance. A number of Crown's subcontracting jobs for the military, many of which were subcontracted from firms like Magnavox and E Systems, required that certain employees have top security clearance to visit other plants and observe processes. As many as fifteen employees had such clearance.

In the early sixties, Crown also made a smaller stainless steel shock mount to go in the nose of the B-52 bomber. This item was roughly two by five inches and looked like a double squirrel cage; it had to support the plate on which a radar mechanism sat. The stainless steel welding was delicate work that the Navy had to approve. "One of the welds was bad in that shock mount," says Wilbert Will. "We had to learn how to heli-arc them. We chamfered those holes and that's how we got them to work." Crown also made replacement parts for the shock mounts.

The Allison Generator, manufactured by Aeroproducts, was another important contracting job. Lockheed's four-engine Constellation planes used engines with Allison propellers, and in a short period in 1966, two of the planes crashed when taking off from Newark Airport. Many deaths resulted and for a while Newark Airport was shut down. Investigators found an electrical coil assembly in Allison's propellor feathering mechanism to be faulty. Aeroproducts had built the mechanism using a coil assembly made by a Chicago pinball operator, and now asked Crown to redesign it and build a new one. At a dollar a coil more than the Chicago firm had charged, Crown made coils that lasted, using Mylar tape, which was new at the time. Crown had to work seven days a week on the coil assembly because aircraft were grounded, and had to pass a tight inspection to meet A-quality production requirements.

Allison Generators — an Indianapolis firm now called Detroit Diesel Allison — also produced an emergency generator that would drop out of the bottom of a plane in case of power failure; a propeller would then spin to generate power for the plane's control surface motors. For years, Crown built those generators, first for Aeroproducts in Vandalia, Ohio, then for Allison, and then for Marquardt, in California. Crown also manufactured a coil assembly, one of the generator's components.

Crown took on jobs like this because, as Arnie Heitkamp explains, "Nothing really scares you if you have hands-on people. If you want to do it, you *can* do it, and you *do* it." And working together on these projects created a great feeling of camaraderie. "I've worked at other plants and there were always differences between departments," says Bob Marshall. "At Crown we were a pretty close-knit group because we all grew up together. We were all trying to protect our jobs, and we'd get a pay raise once in a while by doing a better job and keeping the company going. It was a great small, young company to work for."

In 1967, the government was desperate to find a supplier of microphone and earphone elements for a self-powered communications device used in combat. Only a few U.S. firms manufactured the parts, and the government had 26,000 requests on back order. Crown had been a reputable supplier for Gentile, so the government asked if they would consider manufacturing the parts, and Crown agreed to produce 41,000 earphone elements and 16,000 microphone parts. Crown had no good audio people on staff, so they brought in Jim Heilers, who saw the project through.

One of the last things Crown made for the government — in the early '70s, in the old Excello Building — was focus coils. The very last thing it made was TA-1 handsets for a battery-operated field phone system for use in Vietnam. For this they used surplus parts from the old Signal Corps telephone system. Over three or four years, Crown made about 50,000 of the TA-1 receivers at about $15 apiece. For a company the size of Crown at that time, orders totalling $750,000 were still a big deal, and would support a lot of lift truck development. ❍

Versatility *Crown's clients included IBM and National Cash Register, for whom it made a card reader, a device to read those old "do not fold, spindle, or mutilate" cards that computers once needed for data processing. Opposite page, card selector made for Access.*

From Novelties to IBM

(1952–90)

6.

The decade of the 1950s was Crown's decade for experimentation. "In a sense," says Jim II, Crown's current president, "Grandpa Dicke's death was the moment at which the company hit bottom, struggled, and started to grow. We explored manufacturing novelty items, and made every mistake you can imagine. We rejected the idea of manufacturing a trash compactor, because who would want to compact their trash? (Remember, this was 1952.) We rejected the

Trial and error

In 1954 and 1955, Crown produced several novelty items, including ice stoppers (above), which it wasn't set up to market. These went to the Boy Scouts, as troop fundraisers.

idea of manufacturing an electric can opener. Why would anybody want one when it was so easy with a wall-mounted crank can opener at a tenth the price?" (Besides, the thinking at the time was that plastics would replace tin cans.)

"Instead, we made ice stoppers that would keep the ice in your glass from bumping you in the nose. We made 'utilitongs' for getting olives out of olive jars." A marvelous item, but too fine perhaps for so modest a function. "We made the Crown saw drill — a combination rotary drill and hand-held jigsaw. I think we sold about 60 of them." Crown had mechanical problems with the saw drills, having bought the engineering from someone else. Crown also made a cable cutter, which presented real engineering problems because the tips that cut the cable kept breaking; the engineers had to come up with an alloy steel to withstand the stress of the job the cutters were supposed to handle. The firm developed and made fishing arrowheads for bow-and-arrow fishermen, the brainstorm of someone who, while out hunting, had seen fish he wished he could catch with his bow and arrow.

In 1954 and 1955, Crown went through a spurt of novelty products that it manufactured and sold to some extent and then dropped, partly because of engineering problems but mostly because of marketing decisions. For the lift trucks Crown's engineers were developing, they were going to have to develop marketing capabilities in material handling. Fishing arrowheads would have to be sold through sporting goods outlets, ice stoppers and utilitongs through kitchen or variety outlets, and cable cutters and saw drills through hardware or tool outlets, and the cost of developing so many markets would have been overwhelming. Jim made the sensible decision to drop the novelty lines. In the end, Crown gave most of the ice stoppers to the Boy Scouts, to sell as troop fundraisers.

The novelty items were not the only products Crown toyed with during the transition from heat regulators and antenna rotators to lift trucks. For a while, as a subcontractor to Joyce-Cridland, they considered manufacturing a hydraulic journal jack to jack up railroad cars so the wheel bearings (or "journal bearings") could be changed, because they would burn out if they weren't properly treated. Crown also purchased the Phoenix Caster Company and all of its inventory. Phoenix had a loyal customer base for its cast-iron casters, and Crown got the customer lists along with the business. Jim Uetrecht remembers going down to Miamisburg, Ohio, hauling inventory to New Bremen and inventorying it, re-engineering it, and establishing a new product line, with the help of two or three other employees. Crown manufactured casters and filled orders, using up the inventory they had bought and ordering enough outside parts to complete the line. But after about a year and a half the Dickes decided they didn't want to be in that business and sold it to the G&W Tool Company in Fort Loramie, Ohio. Some of Crown's branches still carry the casters, which customers still request.

In 1956, the year Crown shipped its first lift truck, Jim Dicke and Verlin Hirschfeld went down to Gentile Depot[1] to discuss the possibility of Crown repairing government equipment under its IRAN program ("Inspect and Repair As Necessary"). They returned with a contract. Jim called Jack Lawler and Paul Muter into his office and said, "When this stuff comes in each item will have a tag on it that tells you what's wrong with it."

"How hard can this job be?" they thought. Then the first batch of equipment came in. Everything had a tag on it, all right, and every tag said: "Don't work. Needs repair." Thus was Crown's Repair Services Division launched.

At first Bob Marshall and Ernie Jameson handled the repair work and Arnie Heitkamp oversaw contract manufacturing. After Gentile shipped equipment into New Bremen, Jack Lawler and Verlin Hirschfeld would analyze it and decide what it would probably take to fix it. Jack would submit prices to the government and get approval to repair or rebuild the machinery or equipment. Most repair jobs involved problems of electrical wiring. Soon Crown had 30 or 40 high-level electronic technicians on its staff, rebuilding jet-engine fuel pumps on a major contract that lasted for years. "They would ship them in here," says one old-timer. "We would tear them down, rebuild them, retest them, and ship them back."

The old mill *Needing space, Crown rented an old flour mill across from Plant 2, to do repair work. Paul Dicke remembers a Marine with a sidearm insisting that Crown improve its security by blackening the high windows in the old mill where the repair technicians were stationed in cubicles. "Man, it's an 18-foot climb up there," Paul said. "It doesn't matter," said the Marine. "Somebody could look in there." Paul painted the windows.*

One repair job was for Magnaflux testing machines (machines used to test metal parts for cracks), machines so heavy that Crown had to rent a cement block building, the old mill building across from Plant 2, to handle them. Even then, when a truck brought the machines, the trucker had to stay until five o'clock so Crown could borrow a lift truck from Stamco (another local manufacturer) to unload them.

Crown rarely said "no" to a customer. One day Jim Dicke called Paul Dicke from Gentile and said, "Can we fix prony brakes?"[2]

"Prony brakes?" asked Paul. "Yeah, I guess so. How big are they?"

"It looks like two of them

come in a case about the size of a desk," said Jim. "The people down here are all stumbling over these prony brakes and trying to find someone to repair and calibrate them."

"Yeah," said Paul. "We ought to be able to fix those."

When Jim got back to Crown, Paul asked, "Do you have any idea what the horsepower was?"

"Sixteen hundred kind of sticks in my mind," Jim replied.

"My God," said Paul, "you'll need a locomotive to test them, you know." But Jim had actually heard 1600 *rpm*, not horsepower, and the prony brakes were relatively easy to repair. "You put new brake pads on, filled tanks with oil or hydraulic fluid, and what have you," recalls Jack Lawler. "It was basically a 100 percent mechanical job that people like Paul Muter and Doyle Birt, who were pretty good mechanics, did a nice job on." The Air Force, unable until then to find anybody to repair the brakes, had simply kept buying new ones and now provided Crown with literally hundreds of them for repair. They arrived by the truckful.

Hands on

Crown got military contract work because employees such as John Koeper, Harold Stammen, and Arnie Heitkamp (from left, in posed photograph) were never afraid to tackle unfamiliar problems. "It just seemed you never knew that it couldn't be done or that it was going to be hard to do."

Another time, Gentile sent Crown's repair service about six oscilloscopes, big electronic units used for testing electronic equipment and "covered with 17,000 knobs and 14,000 dials." The men in repair services were afraid even to plug them in. Jack Lawler reported, "We've got these oscilloscopes in and they're tagged, 'Don't work — Need repair' and we don't have anybody that knows anything about them." Verlin sent Paul Dicke in. Paul plugged one of the big devices in and said, "Good news. It doesn't work, so it's probably just a fuse." They popped open the fuse box and put a new fuse in and that didn't do it. They put a tag on the thing: "Probably Loose Wires."

John Koeper was due in over the weekend. "John was just a kid going to school at the time," says Lawler. "He was in electronics but he was greener than grass. John did figure out how to get the oscillator out of its case, but that was about the extent of it. We ended up boxing those things up and sending them to the people who manufactured them to fix. He could probably repair it today with his eyes shut."

Don't work, needs repair

In 1956, Crown launched its Repair Services Division. It got a contract with Gentile Depot, a government procurement agency near Dayton, to repair military equipment under the government's IRAN program ("Inspect and Repair As Necessary"). Material to be repaired was supposed to arrive with a tag on it identifying the problem. All the tags on the first batch of equipment said simply, "Don't work. Needs repair."

Some of the repair and rebuilding jobs represented substantial work. One of the biggest of these involved the defense warning system known as the DEW line, which stretched across Canada to track enemy aircraft. At each site, a big radar antenna emitted a huge amount of radar energy. Energy was fed to the antenna through a tubular wave guide. Guiding the energy into the wave guide was a focus coil, about a foot wide and two feet high, with a hole through the center that held the neck of the large magnetron tube that was the source of the radar energy. These coils functioned at very high intensity, doing heavy magnetic work to keep the column of energy focused. They were water-cooled but operated close to design limits; the insulation would break down and they would short out, causing a distortion (or loss of radar return) on the radar screen. They had a fairly short life and had to be replaced frequently, which apparently represented a considerable profit to the firm that made them. Gentile began sending the bad focus coils to Crown to rebuild on the IRAN contract. It took the engineers a while to figure out how to rewind the focus coils. The problem was, they didn't know what they were supposed to do, and they had no equipment to test them to see if they functioned properly. Only after Crown had been repairing them for several years did they find out the coils needed to be aligned, and they learned that only after they got an unsatisfactory-repair report, because one of the wave guide tubes could not be focused. Still, Crown cornered a piece of the market on rebuilt focus coils.

What possessed Crown to take on contract work on products they had never worked on before? "I think all of us always felt we were qualified to do anything," says Arnie Heitkamp. "Just give it to us. Crown people are all pretty hands-on people. There wasn't anything that anybody was afraid to tackle."

In those days Crown didn't have engineers with four-year degrees from technical schools. It just had a lot of employees who had a knack for fixing things. "I was always impressed by how

Crown could take something that was totally new and different, something they had never seen before, yet our guys would get some manuals, read them, and the next day be tearing them down and fixing them," says Bob Nieberding, who worked in Sales. "Looking back over those years it just seemed like you never knew that it couldn't be done or that it was going to be hard to do."

At times there were two or three semi-trailers on the way from Gentile with items to be repaired. One semi might contain two Magnaflux machines, fifty prony brakes, or a hundred oscilloscopes, not to mention potentiometers and altimeters and three-feet-high mercury barometers. When Crown got something that nobody on staff knew how to fix they would usually hire someone who could work in that area — if not full time, as a subcontractor. Basically Crown had a reputation for doing good work and for finding people who knew how to make things happen or get them fixed. After Crown's experience with the oscilloscopes, the firm hired some television technicians — Dean Lehman, Bob Vallo, and Jim Feldwisch — to repair the electronic equipment Crown employees didn't know how to fix.

When the Gentile repair work came along, Crown had already started developing the lift truck that would become its major product, although they didn't know that at the time. Why would they take on so much repair work when they had a product to develop? Because it costs money to develop a product, and when you are starting you don't have money coming in; people aren't buying it yet. Contract work gave Crown the income to support lift truck development.

Frankly, they didn't have much choice. If they didn't work on *something* they wouldn't get paid. "When we first started at Crown," says Bob Marshall, "it was a day-to-day operation. If the checks didn't come in, you didn't even know if you were going to meet the payroll. The main object was to get the product out so we could get the money back in. And somehow or another, it worked. In accounting, Adrian Moeller wrote letter after letter, time after time, to our suppliers, telling them, 'Well, we'll give you so much money in thirty days, so much in forty-five days,' and spread it out. He kept talking to our suppliers so they wouldn't cut us off and put us out of business. That to me was a hard job, putting off the bill collectors." Moeller spent many a sleepless night figuring out who to pay and who to put off for a while.

There were periodic cutbacks, layoffs, and pay cuts in the 1950s. "One time I was called up to Jim Sr.'s office," Harold Stammen recalls. "Jim said, 'Harold, I have some good news and some bad news. The bad news is this: Business is slow and it's necessary to make some adjustments in pay and for that reason we're going to ask you to take a reduction of five dollars a week. That's the bad news. The good news is that this is less than the cut I am asking anybody else to take but that's because you make less than anybody else." Jim was smiling when he said it.

"During one slow period — probably in 1955–56 — Crown felt the need to cut back and several people in engineering were laid off," says Stammen. "At first I felt good that I had not been laid off; then a couple of days later I heard they had gotten jobs at Stamco and I thought, 'Those lucky dogs,' because I didn't think I had the experience to work there. There was greater prestige working at Stamco, an older company in town, which had maybe 35 people working in engineering — a lot of mature people with many years' experience and an impressive product. Stamco produced steel mill equipment used for slitting steel. They also made press brakes for bending steel — heavy-duty equipment. In fact, some of the people Stamco hired had less experience than I did. I'm sure glad I didn't go there because Crown became a very successful company."

Crown got its first contract work with the government in 1950; twenty-four years later, it gave up government work, partly because of the hassles and partly because of the low margin. "We started to get into situations where the government was dictating what you would do and how you would do it, and it became impossible to work with them," says Jim Sr. "Not individuals so much as the general policies. It just wasn't worth it, and we finally said we wouldn't take any more contracts."

No fouls

Shirley Schroluke tests foul detectors used in AMF bowling alleys, so bowlers (and their opponents) know when they've stepped over the foul line. One of Crown's salesmen picked this contract up talking to a fellow passenger on a long train trip.

Crown had developed a first-class antenna rotator, but with the rotator essentially what it needed was more quantity, not quality. In Contract Services, Crown developed a tremendous reputation for quality. In 1963, Crown combined the government work and Gentile repair services in one division: Contract Services. Then Ernie Jameson (who had been in charge of repair services) started on the road as a salesman, getting orders from Magnavox, Allison, Baldwin Piano, and other firms in the private sector. Another member of the sales force who played a vital role in bringing in work was Tom Shelby, the sales manager who had made the first contact with Channel Master. "Only someone like Tom could get a job like a foul detector for a bowling alley," says Bob Marshall. "You know how you're not supposed to cross over the line when you're throwing the ball in the bowling alley? There is a little mechanism to detect when your toe goes across the line to indicate you fouled. Tom Shelby was on a train going from somewhere to somewhere else and he got to talking with a guy who worked for American Machine and Foundry (AMF), a big bowling alley firm. This fellow was looking for someone who could build a foul detector for AMF's pinspotter division and Tom said, 'We'll do it.' So they sent all the prints down and we quoted the thing and we got the job." Tom Shelby figured in a lot of stories that way.

Through a Crown sales rep, Fred Stork, they started doing work for Baldwin Piano, of Cincinnati, in May 1963. "Fred sold everybody when he walked in," says Arnie Heitkamp. "He sold die casting, he sold precision, he sold Crown; he sold anything you could make a dollar on. The buyers got tired of seeing him. If you keep going back, they'll give you a print sooner or later" (meaning a blueprint from which you make a bid to manufacture something). "If you keep going back the buyer is going to say, 'Gee, this guy must be serious. This is the fifteenth time he's been here. I'm going to give him a print just to get him off my back.'" Finally Baldwin Piano gave Fred a print and asked for a quote on 10,000 machined parts, so he was happy. He finally had something to take back and get a quote on. Crown started out by doing a die casting of a little round part for the sound system in their piano. Nobody remembers what the part was or how it worked but they machined a lot of them.

The same persistence that landed Baldwin finally landed IBM, too. IBM's buyer sent Fred some prints to keep him from bugging them, and in 1964 Crown made a die casting for the belt drive on the old IBM dictaphone recorder for secretaries. From there they went on to manufacture several other products for IBM. For a while, Crown had a hundred people, mostly women, working on an assembly line making daisy print wheels for IBM's Correcting Selectric typewriter. IBM liked the quality of Crown's work — they shipped to IBM with practically 100 percent quality — and having Crown do it saved them money.

IBM's Lexington plant couldn't get production up to more than 200 daisy wheels a day and they believed Crown couldn't either. So when Crown began producing and shipping 1700 daisy wheels a day, IBM's engineers and quality people spent a lot of time in New Bremen trying to figure out how Crown was able not only to build quality parts but to keep gearing up to that speed. "IBM needed those quantities," says Jack Lawler, "and by golly, every day they were there! They couldn't figure it out."

"They were flabbergasted that we were able to do everything they asked us to do," says Arnie. The fact is, Crown was already geared up to building the antenna rotators and that formula worked pretty much the same way for the daisy wheel. Six weeks of training and Crown was in full production on this competitive, high-production job. Initially, IBM furnished all the parts. Crown was then going to start buying all the parts and do the final machining themselves. In fact Crown pulled the wall off Plant 3 and poured a slab of concrete for a machining center. "There's a hell of a slab of concrete over there still," says Jack Lawler. But Crown soon discovered that it wasn't easy to get good parts.

IBM had a system of qualified and approved vendors: Only when parts suppliers had provided three shipments within ninety days with no rejections were they considered qualified, and only a small percentage of IBM's suppliers ever reached approved status (with parts manufactured to exact specifications, or "to print"). It was possible to maintain a 99.9 percent acceptable level of assembly quality with only the small percentage of parts that were approved, and in the end Crown decided against taking over the buying of parts because it didn't want to pick up the quality problem. Crown assembled IBM's daisy wheels for over a year, then got into pricing problems, and IBM decided to pull out.

Still, Crown's IBM business grew so much that Crown finally hired its own salesmen and had a sales manager for that division, and Stork and Associates no longer represented Crown. Crown reps would spend at least a day a week at IBM. When IBM was making copier machines, Crown made the magnetic pickup roll and the brush roll that puts toner on the paper. Crown engineers came up with the magnetics and helped IBM's engineering people develop the copier roll.

"That's when we decided we needed engineering help," says Jack Lawler "and that's when Neil Good got into the act. They gave us Neil, who was also working rotators in the early seventies." Neil died in 1984 and Bill Harshbarger took over.

Neil had incredible rapport with IBM. "IBM thought Neil could do no wrong," says John Koeper. "If Bill at IBM was writing with a blue pencil, Neil could say, 'You've got a nice red pencil there, Bill.' Bill would say, 'No it's blue,' and Neil would say, 'No, it's red.' Bill would say, 'God, I must be color blind, Neil. You're right, it's red.' Usually what Neil said was the way it came out. Neil did a whale of a job. Of course, Bill did a good job too. They both did."

"At that time we needed more than occasional engineering support," says Lawler. "We needed an engineer to oversee the work. A lot of the engineers who later worked on the lift trucks got started in the subcontract work in the sixties and early seventies."

"And the guiding spirit over everything was Jim Dicke," says Marshall. "He was overseeing the whole operation and seeing that everything was running as smoothly as it could with what we had to work with. He just made things happen. He had a sense of how to get things done, and of course he had good help in Warren Webster."

After Crown's reputation for quality spread, jobs came in from unexpected sources. Fred Stork had even landed a contract with National Cash Register, making Crown the first outside engineering firm to design and make a product for NCR. NCR had a reputation for engineering all its own stuff in an ivory tower atmosphere, and its management had begun to wonder why it took them umpteen years to develop a new product when IBM cranked them out like crazy. "That's because we do everything ourselves at our own rate," someone at NCR said, so the firm decided to try an outside vendor and came to Crown.

Crown designed a Computronic card reader, a device that read IBM cards (the old "do not fold, spindle, or mutilate" cards computers once needed for data input). The card reader took a card through a slot, sensed the location of the holes punched in it, then read them out in conjunction with a business machine. Crown also manufactured the System 60 card reader for Access Corporation (in Cincinnati) and sold some card readers to Hickok Electronics in Cleveland. Soon, of course, IBM cards would be a relic.

For a while Crown had done a fair amount of work for Magnavox, then that trailed off. They did some work for Ecor and E Systems but after a while, says Jack Lawler, "Crown tried to put all its subcontracting eggs in one basket with IBM." They started with IBM's Lexington plant, then went to Boulder, Tucson, and San Jose. They even did a little work in Boca Raton, when IBM got into robotics. Crown got that order partly because Crown had an IBM robot. Someone at IBM's Charlotte, North Carolina plant said, "I think it would be nice if you could say all this welding was done on an IBM robot, and Crown has a robot, so maybe Crown ought to have the orders." IBM sent Crown the order to make all the support braces (weldments) for a large robot IBM manufactured in its Boca Raton, Florida plant. Crown produced the weldments but did so using the robot of an IBM competitor.

Selling Crown

Ernie Jameson sold to Magnavox, in whose buying office several buyers shared one telephone on a post with a telescoping scissors mechanism. If one needed the phone, it was swung around for him.

In 1989, Crown started pulling out of contract work. In April 1990, Crown finished its last project for IBM, making parts for an IBM cash machine for IBM's Charlotte, North Carolina, facility. Then Crown said goodbye to contract work. The material handling division had gotten so big and was running so tight on plant space that top management said, "We've got to put our efforts toward the lift trucks now. Contract work served its purpose but we've got to move on."

"Lots of times while we were doing the subcontract work," says Bob Marshall, "I thought, if we spent as much time on our lift trucks, getting them right and working on the product, as we do on other people's products, we would have the best lift truck in the world — not realizing at the time that we needed the money that all these other little projects were funneling in to us in order to make the lift trucks possible. Jim knew what he was doing. I didn't know it. I'm sure Verlin did, and some others who were closer to the contract work: Ernie Jameson and maybe Paul Dicke, Jim Sr.'s cousin."

Forklift manufacturing is heavy manufacturing that involves moving parts. It requires machining, and machine tools are very expensive. "Subcontracting helped there," says Tom Bidwell. "We would buy a machine that we needed to fulfill a subcontract order, and we would use that same machine to make a lift truck part." Sometimes the people who gave Crown the contract helped pay for the machine and sometimes Crown would buy it based on a large order they were getting, thinking, at the end of this order we will have this machine half (or completely) paid for and then we will make our lift truck parts on it. When Crown bid on a job, they would think, 'What machine could we buy for the plant?' If they won the right job they could maybe buy a $5000 machine, try to pay for it with one order, and when that order was done they could make lift truck parts. Or they might run it eight hours on a contract job and eight hours on lift trucks. So the contract work helped cover overhead costs and supply new machine capabilities.

Part of Jim Sr.'s genius was knowing what to do with the numbers others gave him. Paul Dicke would tell Jim and Ernie Jameson that he could make a certain product for a dollar, and Jim would say, "That may be so, but we can get ten dollars for it because the nearest one of the same general variety is selling for eighteen dollars."

By helping to finance the purchase of machine tools, contract work (together with the antenna rotator) helped subsidize the development of Crown's lift truck — helped Crown keep reinvesting in its new product line. Contract work kept the machines busy during slow times and provided a way to buy more equipment faster. Eventually, of course, Crown got so busy that the machines were completely occupied making lift trucks. If filling a contract order meant they had to stop making lift truck parts, it no longer made good financial sense.

Crown's strength had been to produce small gears, transformers, and motors, first for heat regulators and then for antenna rotators. Now contract work had given Crown experience and know-how in many areas it hadn't known before. The experience was particularly invaluable for Crown's degreeless young engineers, who had gone no further than high school but who were pragmatic and mechanically inclined. Each time Crown was asked to quote something or to build a product to which they had never been exposed, they were learning things that would come in handy later. Working on a subcontract item, they would often think "Why don't we do that on a lift truck?" The contract and subcontract work helped to meet overhead with a little extra, provided the cash to buy new machinery, and financed incredible and varied on-the-job training for Crown's technical workers and engineers. Crown got better and better at manufacturing during those two decades of contract work. Contracting and the antenna rotator together subsidized development of the lift truck and gave Crown a tremendous start — and it grew from there. ❍

[1] *Gentile Depot was located on Wilmington Pike in Kettering, Ohio, a suburb of Dayton, near Wright Field. Named for Major Don Gentile, a highly decorated World War II Air Force pilot from Piqua, Ohio, it later became the Defense Electronics Supply Center, which Daytonians soon shortened to DESC (pronounced "DESSY"). Patterson Field and the adjacent Wright Field later became Wright-Patterson Air Force Base.*

[2] *A prony (or "pony") brake is a friction brake or absorption dynamometer, a testing device to measure horsepower. To use a prony brake — a big round object with a spring balance (or a weighted lever) and a brake mechanism — you run an engine at a certain speed and keep increasing braking force until the speed can no longer be maintained. The spring balance tells you what the force was, and a simple computation using rpm and the diameter of the braking surface gives you the horsepower.*

Drawing board, yes. Desk, no. *Crown management saw desks as a sure sign workers were pure overhead. A drafting board was something else, because you were producing a drawing. From left, John Koeper, Al Schwieterman, Duane Hegemier, Jack Lawler, Jim Uetrecht, and Harold Stammen (standing in back, Ernie Jameson). Facing page, Verlin Hirschfeld.*

A Chance *to* Prove Themselves

(1945–51)

7.

"In the early days hiring was tough," says Verlin Hirschfeld, Crown's retired general manager. "Now it's easy. Everybody wants to come work at Crown. But in the early days we had Goodyear seven miles down the road, a big company paying very well, and we had other companies around with established products. For us to compete was tough because everyone knew we were struggling and their careers might be short. If they wanted job security they didn't come to Crown. At that time I think I was doing most of the hiring. As general manager back then, you were just general: a little bit of everything and not an expert at anything. But I've lived here all my life, and being in the retail business, I knew everybody and they knew me, and people did come talk to us about coming to work. I think we always had a pretty good feeling whether they were or were not the right kind of employee. Those who came came because they thought Crown just might make it and there might be an opportunity here.

"Jim made it clear from the start that Crown was a private company, that it would stay a private company, and that there would never be a share of stock sold to anybody," says Verlin. If you came to work in management, if Crown did well you would do well — on salary and bonuses — but you would not own part of the company. "And Jim stuck to his word. I never got one share of stock, but they took care of me," says Verlin.

"Crown is like the old Pittsburgh Pirates," says Arnie Heitkamp. "It's a family-type company. You feel comfortable talking to people. If you have ideas you're not afraid to bring them up, whereas in bigger companies a lot of people don't, for fear of whatever. Maybe it had something to do with the fact that Crown didn't have college-educated people. The people at Crown, if they were educated they educated themselves, like myself. I never went to college, I took ICS correspondence. I did take purchasing courses back when I ran purchasing and inventory control."

No complaints

For years, Charlie Zwiebel worked alone, making special lifts. "I was danged near my own boss," he said. When he retired, perhaps fifteen people were customizing lifts.

Crown's willingness to hire people without a college education has produced some very loyal employees. "If I had been in Dayton," says Wilbert Will, who supervised Crown's tool development, "I would have been operating a machine in a tool and die shop, I'm sure. I probably wouldn't be in a management job because I never went to college. I just evolved into this job myself, here at Crown."

"Crown didn't seem to have any preconceived ideas of what you could or couldn't do," says engineer Jerry Pulskamp. "They would give you a shot, and if you could do your job, you had the job."

"They didn't hold you back because you were young *or* because you were old," says Dennis Dicke, a distant cousin of Jim's who was made supervisor of shipping and receiving at the age of 28. "They liked youth. They gave them the chance to prove themselves."

In the company's early days, there was general skepticism about the value of a college education. Few people at Crown had one — maybe this was true in manufacturing firms all over the country — and there was a sense that too much "book learning" made people lose touch with the real world. Instead, Jim encouraged people to learn on their own. He didn't have that faith in college that became so important later.[1] "Look at Jim Uetrecht, Harold Stammen, and Paul Dicke," says Wilbert Will. "I don't know any of them who had previous experience when they got into engineering. They just learned it on their own. We're so small and work with so few things that you don't have to have a wide knowledge of everything. You just accumulate it. You learn on the job."

Paul Dicke had two years of mechanical engineering at the University of Vermont, but the rest of his education was fairly practical. One of the things college taught him was where and how to look things up, because much of engineering is technical data found in books. Efficiency is so important these days that few firms will hire an engineer without college credentials, but as Jerry Pulskamp put it, "If you look back in the history of our country, a lot of awesome engineering was done by a lot of people who didn't even have a high school education."

One of Crown's strengths has also been patience about matching the right person with the right job, even if it takes years to figure out what that match might be. If an employee doesn't work out in one department, Crown will usually move them somewhere else until they do work out. "You rarely hear that somebody was let go because they just didn't work out," says Wayne Schroer.

Sometimes Crown's flexibility was unnerving — almost as if it were a conscious effort to keep you on your toes. "It would be many years before you got a desk," says Harold Stammen, "even if you were doing the kind of work that involved only a desk. I believe something about giving a person a desk bothered top management. They saw it as a sure sign that you were pure overhead. Putting somebody on a drafting board was something else, because you were producing a drawing, something you could tangibly see. I assumed when they decided to get me a desk that it was probably because they needed my drawing board for somebody else.

"Along with that," says Harold, "titles were not readily handed out either — and there was good reason for that. It allowed flexibility on moving responsibilities to the most capable person, as the need arose. You don't want people getting hung up on titles, was the philosophy. It tends to create boundaries and sometimes barriers to getting the job done."

Nor did Jim want to see a lot of paperwork; he considered it a sign of inefficiency. Among other symptoms of inefficiency that he scorned were three-ring binders, especially new ones. In the early years, Crown bought mostly used furniture and used equipment. "Every time I go to one of these sales," Jim would say, "they seem to be loaded up with three-ring binders. That's a sign of poor business." So Crown did not buy many. Jim wanted production, not reports. ❍

[1] *Later, Jim Sr. and John Herkenhoff helped start the Western Ohio Educational Foundation. Jim was also on the board when Wright State University opened its Lake Campus branch. Years later his son Dane served on the board, and then Jim II. Jim III's wife Katy serves on it now.*

The Bumper-Upper *After the war, Crown manufactured Joyce-Cridland's auto jack, the Bumper-Upper (shown here, the Power Jack, a later powered version). Sales fizzled, but Crown incorporated the pump from the jack into an adjustable-height lift table for the tool room.*

Crown Breaks *into the* Lift Truck Market

(1956–72)

8.

The value of the subcontracting experience became apparent as Crown developed its lift truck, but one contract in particular led directly to Crown's new product. In the early to mid-1950s, Crown became a subcontractor to Joyce-Cridland, a Dayton firm that manufactured hydraulic lifts for lifting cars, trucks, and busses so mechanics could service them from below. Jim Dicke's father-in-law, Warren Webster, was one of Joyce's co-owners, and as Jim's mentor he well knew how the quality of Crown's manufacturing had improved since the early days. Jim's relationship with Webster would play a significant role in Crown's evolution.

Among Joyce's product lines was a hydraulic hand jack (a "speed-jack") used in industry and construction. Joyce had been subcontracting production on the jack to a Canadian licensee, the Midland Foundry & Machine Company of Ontario, which manufactured and marketed a number of Joyce-designed jacks and lifts in Canada and manufactured some that were marketed under Joyce's name in the United States. The quality of production at the Canadian firm left much to be desired. The firm's engineering drawings were inadequate and the firm had no inspection process, so all of the parts made one day would work together, but the parts made one day wouldn't necessarily work with the parts made *another* day. And management's idea of good employee relationships was to have the janitor come in an hour early in freezing weather to turn the power on so the coolant used in the lathes wouldn't be so hard on the workers' hands. Webster asked Jim if Crown would take over production of the hydraulic hand jacks sold in the United States. With Joyce's approval, Crown reworked the engineering drawings and began producing Joyce's "Yello-Jackit Liftmaster Jack." Available in capacities from three to one hundred tons, the Yello-Jackit could handle a range of jobs, from lifting light machinery to supporting construction in mines, mills, and factories.

Then Joyce (which marketed primarily to the automotive business) decided to develop a bumper jack, and Webster asked Jim to help Joyce manufacture a redesigned hydraulic hand jack for lifting autos. That jack evolved into a model Joyce called the Bumper-Upper. The Bumper-Upper, which you hand pumped, could lift a car by either the front or rear bumper. Crown worked on the design and manufactured the jack, plus some of the other components, to cut costs. Thieman Stamping made the frame. In 1959, Crown introduced a battery-powered version of the Bumper-Upper called the Power Jack. But before the first 200 units had been assembled, another company introduced an air-powered unit that changed the paradigm for service-station bumper-jack design. The air-powered jack, which was powered by the compressed air readily available in service stations, quickly became the favored design because of its simplicity and operating speed. In any case, American cars had evolved to a stage in which car bumpers were no longer rigid enough to lift a ton or so of car without sustaining damage to the car. Jim decided to stop production of the Power Jack at 200, and the Bumper-Upper was also eventually phased out.

The Bumper-Upper never amounted to much in sales, but in an odd way it was a key to Crown's entering the material handling industry. Crown would soon incorporate part of the Bumper-Upper into an adjustable-height lift table for the toolroom environment. Meanwhile, co-designing Joyce jacks had given Crown valuable know-how in heavy lifting.

Channel Master's offer to take over the worldwide marketing of Crown's antenna rotators had been an offer Jim couldn't turn down, but it left Crown with nothing to market. One day, Jim was sitting at the kitchen table in his father-in-law's house in

Kettering, Ohio (a suburb of Dayton), telling Webster his problem. It so happened that Webster had been thinking about a product that he felt needed to be developed, a product for which he felt there might be a market. At Joyce-Cridland, when tools and dies had to be changed on machinery, someone had to lift the heavy metal pieces out of the firm's small stock room, raise them to the level of the machinery, install them, and eventually reverse the process and return them to storage. In one three-day period, two different workers had hurt themselves lifting the heavy tooling. One had hurt his back, and the other had been taken by ambulance for hernia surgery. "I wish we had a lift that could raise the dies up and put them on a shelf," Webster had remarked to Clarence Bradley (whose machine shop in Centerville got most of its work from Joyce). Webster wasn't concerned only about injuries to the workers who had to handle heavy, bulky objects; those production dies were expensive, and dropping one could stop production in the whole company.

One day Webster was buying filters in W. H. Kiefaber's, a Dayton plumbing supply company, when he saw two small hydraulic hand pump lifts with a 500-pound lifting capacity — a model called Big Joe. "Those darn lifts were just about what I thought we needed to lift these structures onto the mills," said Webster, "but there was a puddle of oil on the floor running out of the Big Joe lifts because the seals hadn't held. I thought, 'We ought to be able to build one better than that.'" Webster figured it would be easy to improve on the Big Joe, to manufacture a "lift table" that could be used to raise and lower heavy items from floor to shelf or table height and back. Webster wasn't even sure

Basic devices for lifting

In the material handling industry various devices, incorporating pulleys or levers, are used to help lift and move raw materials, products, and parts.

A simple ***dolly***, a platform on wheels, enables people to move small but heavy loads around easily.

A ***jack*** — for example, an automobile jack used to change tires — is a mechanical device used to lift heavy weights a short distance. A jack can be portable or stationary. Add a motor, a brake, even a computerized control and a jack may still be only a jack — a device to raise and lower a load.

But add a platform on which to mount heavy objects and you have a ***lift***.

If you add wheels to the lift, so that it is not stationary, you have a ***lift truck***. Most lift trucks are based on the principle of the ***hydraulic jack***. Hydraulic jacks use the force created by liquid under pressure to greatly augment the mechanical force exerted by an operator. On ***powered lifts***, the jack is replaced by a hydraulic cylinder and a motor-driven pump.

if his two partners at Joyce would want to build this thing, and as it turned out, they didn't; one of them called it a fly-by-night idea. But Webster thought the device was a good idea because it would reduce labor and increase safety. When he mentioned the idea to his son-in-law, Jim said, "Sounds good to me."

Jointly, Crown and Joyce developed a design for what they were calling a lift table, and gave the drawings to Clarence Bradley, whose Centerville Steel Products Co. built the first, rough prototype. They had taken portions of the design of the Bumper-Upper and made it into a lift table. To the existing hydraulic unit, they added a chain and pulley. For a frame, they put some uprights on a base, and added wheels and casters and a flat surface to the elevating carriage. At first the men couldn't figure out how to design the pump and hydraulic lift cylinder to raise the lift. "That doesn't seem like a big job today," said Webster, "but it was a major operation then." Finally, it occurred to Webster that the hydraulic unit from Joyce's Bumper-Upper could provide the pumping and lift action for the new lift.

That connection might seem obvious now but it wasn't at the time. It didn't seem, to the men, that to lift objects weighing only several hundred pounds they would need a hydraulic cylinder with a diameter as wide as the cylinder on the Bumper-Upper, which was designed to lift a car weighing a ton or more. A larger diameter would be more expensive to produce, and at first they thought they could get by with a cylinder with a smaller diameter. Eventually they realized that although the lift table would be lifting lighter loads, it would be lifting them considerably higher. The Bumper-Upper lifted loads only 30 inches or so — the height of a vehicle's bumper. The lift table was going to be lifting loads twice as high, so it would need about the same size hydraulic cylinder, with a pulley and chain mechanism to provide a 2:1 lift ratio.

In the end, the tool-and-die table simply went a little further than the Bumper-Upper. Instead of bumper rests for grabbing the car bumper it had a flat platform on which to set and lift the die, which was secured to the plate with straps. Pushing the lift table was like pushing a handcart except that it could also lift 500 pounds. Of course, pushing it after it held 500 pounds was another matter. And all the lift table did was lift and lower the die; the worker had to manually remove the die from the lift and shove it into position.

Joyce-Cridland marketed that first lift table as the Joyce "Material-Handler," the PLT 5 (500-pound capacity) and the PLT 10 (1,000-pound capacity). It did not sell well; moreover, there were design problems. For one thing, the casters were out front and the load wheels in back, so when you put a load on the platform, the casters froze up; with too much weight on the casters, you couldn't steer or swivel the lift table. Even more troublesome, the platform was not stable enough, and the gusset was positioned in such a way that if you had your hand in the wrong place your finger could be hurt when the platform came down — as you were lowering something.

Jim asked Tom Bidwell, a quiet young engineer the firm had just hired, to design a new prototype model. In painstaking detail, and using some of the same components from that first, unstable prototype, Tom redesigned the lift table, putting the swivel wheels on the back and the straight load wheels out front. He convinced Webster, a tall man, that the push-handle at the top of the columns was too high for most people, and lowered the handle. He made sketches, design layouts, and parts drawings of his improved design.

"My first project at Crown was to design this little lift table," says Bidwell. "I've been doing it ever since." Although they called it a hydraulic lift table while they were working on it, that first Bidwell design became Crown's first hand-pushed, hand-pumped stacker, the LT500 (lift truck, 500-pound capacity). Next he designed the 1000-pound capacity LT1000, the FL1000 (with an adjustable fork), and other trucks with special features.

Niche player

Walter Rump puts die on a punch press using the LT500 lift truck, which allowed Crown to enter a niche in the material handling industry that had not interested major players.

After completing the basic design, Tom spent hours refining the side, front, and back views and coloring the unit in various color schemes. In those early meetings about the lift truck the men discussed what color to paint it. Tom wanted to paint it a dark red. "Well, I don't," said Jim. "You can paint it any color you want, but we have a lot of blue and silver paint in inventory." The first trucks were blue and silver. As the designs developed, Jim began stopping so often at Tom's drafting table that they installed a telephone so Jim's talking with Tom wouldn't interfere with other engineering operations.

Meanwhile, Webster's partners at Joyce-Cridland decided they didn't want to manufacture and sell the new lift table, because it wasn't geared to the automotive market to which their sales were directed. And that made sense; the product would be used primarily in warehouses and factories. Of course, Crown had been selling heat regulators and television antenna rotators, so the lift table wasn't exactly something their salesmen knew how to sell either. But that didn't stop Crown, which knew it had to develop a new market anyway. Was it a big change in direction, from selling an antenna rotator to selling a lift table? "Not really," says Crown's Bob Nieberding. "Selling is selling. You've got to know your product, you've got to know your market. But selling is selling."

First, of course, you have to identify potential customers, and Crown now had to find a new kind of customer. In a way, that was a lot of fun, says Nieberding. "When you look back over those years it just seemed like you never knew that it couldn't be done or that it was going to be hard to do. You never thought of it that way. You'd just say, 'Well, here's something new for us to do,' and you'd jump on it.' Sure, you'd make mistakes as you went along, but you learned from the mistakes, and the next thing you knew we were setting up dealers in St. Louis, Detroit, Cincinnati, and New York to sell our lift trucks."

Crown shipped its first lift table in 1956; now they had to figure out how to sell the thing. Crown's main competitor was clearly the Big Joe. In 1957 Crown introduced the BL (battery lift) truck at 1,250-pound capacity, 250 pounds heavier than Big Joe. Big Joe increased its capacity to 1500 pounds, which Crown then matched. By 1959, Big Joe had a $299 price on a 500-pound-capacity truck much like one Crown had to price at $439 to make any money. Everyone said, "How can you charge $439 for the Crown lift when there's only six inches difference in the lifting height? I can buy a Big Joe for $299." Jim Dicke and Tom Bidwell pondered: How on earth did Big Joe do it for that price?

That year, Crown had nearly completed the design on, and was planning to introduce, its new H (hand-operated) and B (battery-operated) series, to replace the LT, FL, and BL units. Crown had been studying Big Joe's pricing and one day Tom Bidwell figured it out: "They don't *sell* any of them — because they won't reach the back of the tailgate on a truck. If you order it ten inches higher, which is the next model increment, that ten inches costs one hundred dollars more." So Crown just went completely around Big Joe, introducing a lift truck with a different model number at $297, and at $297, Crown got its foot in the door. The accounting department said, "You're going to lose money on every $297 lift." Jim and Tom responded, "That's right, except we're not going to sell many of them." And they didn't. "That's the way it worked," says Jim, "and that's just paying attention."

"Our biggest competitor was Big Joe," says Nieberding. "And I always understood that they got started by going to Chicago and literally selling out of a booth at a trade show stuff they hadn't even made yet. Then they went back and started building and developing it. They were probably the pioneers in the small lift truck that we started with. They were recognized at that time as probably the leader in these smaller trucks. We followed them for a while, duplicating their model types of equipment, then at some point we quit following and became leaders."

At first Crown sold with literature and with salesmen explaining the benefits of owning such a product: In a small machine shop where there was a lot of strenuous manual work, this relatively inexpensive piece of equipment could handle a die for you — get it off the truck, get it to the shelf, store it, and put it in your press when you needed to run it.

Crown had to decide whether to use its own salespeople or sell through commissioned sales representatives who carried other lines and other products. They started, as Jim recalls, by calling on material handling dealers. Tom Bidwell did a lot of that. "Tom used to fly out of Celina, Ohio, in that one-engine airplane. The biggest problem he had was that he had only one or two pieces of lifting equipment, where Big Joe had three, four, five, or six pieces. They'd say, 'Oh, we got the Big Joe line. We don't want that.' So we had a problem growing at first."

But grow they did because, under Jim Dicke and Tom Bidwell, Crown did a superb job of marketing. They had sales people calling on grocery warehouses, but they also took full advantage of trade shows, where, says Bob Nieberding, "people were looking for certain products and would seek out what they wanted. We had them coming to us."

Most of Crown's participation in the early trade shows was designed to pick up dealers, and they settled for some less than perfect dealers just to get geographic coverage. They had a pretty model in their booth at the trade fairs who would give a prospective dealer a button that said "Handle with Crown," and then, before he could get away, the girl would introduce him to Mr. Bidwell, "the closer." Looking back, it seems corny — but it worked.

"Of course, all the time we were doing this we were learning an awful lot about the industry and what was needed," says Tom Bidwell. "I think the first list I made of competitors, there were something like forty-seven companies on the list. It took a lot of effort just to find out who the forty-seven were — and we were down at the bottom, believe me. We literally started at the bottom and kept expanding and growing and just kept climbing up the list."

After World War II, Clark and Yale were the big firms in the forklift truck industry; during the war they had supplied the government with many of the trucks used for handling cargo. Material handling was an immature industry, however, and Crown entered in a small niche. In a way, Crown's new-boy-on-the-block status was an advantage: Crown did not have to modify existing models to create a new model. Their competitors already had a line of lift trucks, says Jim Dicke, and "they would say to their engineering department, 'Make us a new truck, but use the transmission from this truck, be sure to use the lifting devices from that truck, and standardize on this cylinder because we are using those over on this truck.' They were tying the hands of their engineering department, whereas Crown was starting from scratch. We started and made a family of trucks and worked from there." Thus, early on, Crown developed strength in design and styling. They used that strength to develop the niche in which they entered the industry, and they also developed one of the strongest marketing organizations in their field.

In the early days, they developed a lot of specialty lifts. Soon after they introduced their first hand-operated and battery-operated units, they began getting requests from customers to make specially modified trucks. Over a few years, they developed

Leadership

Jim Dicke Sr. (left) always had someone he could depend on. Here he talks with Tom Bidwell (center) and Verlin Hirschfeld. Tom moved from engineering into marketing and at one point handled both engineering and sales of Crown's lift trucks. Verlin handled the business part of the business and kept a tight rein on the company wallet. Jim motivated employees to do their best, gave them credit when they had a good idea, knew when to tell them to run with it, and had a healthy attitude toward mistakes.

many special configurations, each for a specific customer or type of application.

Crown learned everything the hard way, including how to choose a customer. They designed a truck to transport tobacco leaves that was popular and sold well — but learned belatedly that the tobacco season is only eight weeks long. Customers in the tobacco industry would order the trucks, not pay the invoice, and send the trucks back when the season was over, at about the moment the invoice was past due.

"Back then," says Harold Stammen, "we were a new company and very concerned about getting every order we possibly could to keep production going. I remember making field trips to customers' sites, traveling with Tom Bidwell (who had one foot in engineering and one foot in marketing), seeing what their needs were, talking to them. Shortly after Tom had completed the basic design of these trucks we started to receive orders with requests to modify trucks. That's when I became involved in lift trucks: making special modifications — modifying the lift height, the platform, or the fork carriage. We had some real unusual specials. We had a request to put *ac* motors on some (instead of *dc* motors). I remember designing an explosion-proof *ac*-type lift truck to be used in a gas environment. They wanted power trucks in which any internal electrical sparks emitted would not set off an external explosion."

Among Crown's early specials were stockpickers and order pickers for the U.S. government and a hamper-dumper truck for the U.S. Post Office (which they would use to lift mail hampers and dump the mail from them). Crown had already begun

◀ *Funeral directors wanted their trucks for caskets and cadavers in normal colors. "We are not drab people," they said. "We want colors like everyone else."*

◀ *In 1956, systems analyst Bob Marshall tries out a special lift with hand controls. Bob brought order to production.*

◀ *This hamper dumper was designed for the postal service, which used it to lift heavy mail hampers and dump the mail from them.*

doing contract work for the government; these special trucks gave Crown more experience both with federal contract work and with lift truck development. More important, they helped the firm get a concrete ramp and other testing facilities important to product safety.

Crown also made a lot of special trucks for handling caskets and cadavers. "Those early lift trucks for handling cadavers sort of stick in the memory because they were so unusual and we built more than one of them," says Bob Nieberding. When Tom Bidwell was designing a funeral home truck, one funeral home wanted two rods to telescope outside of the lift platform to lift bodies for placement in the casket. The funeral director Tom Bidwell was working with told Tom he needed two straps.

"Are you sure you don't need a third strap in the middle to keep the body from drooping?" Tom asked.

"No," replied the funeral director, dryly. "That's not a problem. You've heard of stiffs, haven't you? Well, believe me, when they're stiff, they're stiff: Rigor mortis sets in. You don't need any more slings." Crown painted the first funeral trucks bronze because support equipment for a casket always seemed to be bronze, but after the trucks were shipped the funeral directors said, "We would really like to have standard Crown colors, the blue, silver, and a touch of red. We are not drab people. We want colors just like everybody else."

Crown's reputation for doing quality work got them one highly specialized order from Stamco, the local firm from which Crown had occasionally borrowed lift trucks. Stamco, which made

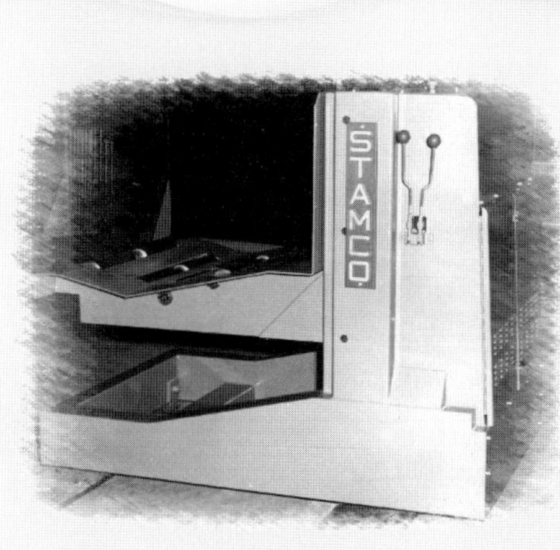

For Stamco, Crown designed this coil buggy to lift rolls of steel weighing as much as 15 tons.

With this popular truck for carrying tobacco leaves, Crown learned to avoid short-season customers. Some customers returned the truck before they paid the invoice.

heavy-duty steel-handling and metalworking equipment (such as slitters and rollers), asked Crown to make it some coil buggies, because it wasn't happy with the coil buggies it was getting from someone else. In the steel mills, a wide sheet of steel goes down what they call a slitter line, where it is sliced into smaller strips. These strips are then rolled into coils, like rolls of paper towels. In some parts of the country you can see these coils of steel going down the highway in tray-rolls on the back of a flatbed truck. Coil buggies in capacities from 10,000 to 30,000 pounds are used to lift and transport the coiled steel along fixed rails. Crown designed Stamco's coil buggies, engineered them, fabricated and assembled the parts, and shipped them to Stamco, which shipped them out to their steel mills.

More often than not, Crown would build a truck for a customer for a specific need, take a picture of it to put in the file, and never build that particular kind of lift truck again. The next day they would be given a totally different set of requirements. "Maybe half of what we built in the early days were specials," says Nieberding. "We did a lot of that in order to get business. A fellow would look at the design and say, 'That platform's only 26 inches square and I need something a little bigger, 30 by 30.' And we'd say, 'No problem; we can build that.' It was more expensive if you customized it — especially if you only built one and the next day you built something different — but it was an important part of our business." It was also expensive: Each model used entirely different main parts and inventory requirements were considerable.

Crown's special-lift department still does customizing, although not as much as it did in the early days. "You prefer not to do them," say the salesmen, "but they are part of doing business." The specials take a lot of time and can interfere with normal production. In the late 1950s, the engineering department was so busy designing and drawing special lifts, as well as supporting models in production, that they had little time for product development. Four or five engineers were working on specials and one or two on product support. They were trying to squeeze in new product design, but at first that was given a lower priority; sometimes no work was done on new product design and development. The emphasis needed to shift.

"The biggest change in those earlier days," says Nieberding, "was the day in 1959 we finally said to ourselves, 'We're doing too much of this; we're getting nowhere. We're spending all of our time building one-of-a-kind trucks.' That was the day we sat down and developed what we call our H series (a hand-operated line of trucks) and our B series (battery-powered lifts). One model had a platform on it, some models had forks, one model had a telescopic mast, and all of these models were available in four different lifting heights. That was a complete line of trucks. All of a sudden we had maybe 30 or 36 standard trucks, so a sales rep could say, 'Well, if *this* standard model doesn't fit your needs what if we go to *this* standard model?' They could guide a customer to another standard model rather than a special design. I feel that's when we really got into material handling, when we made the break from all that specialized equipment and had a standard line."

To limit the number of specials, in other words, Crown

decided to expand its standard offering. Crown totally redesigned the truck so that it was more rugged and so there was more commonality of parts among the models. Interchangeability of parts made production simpler and reduced the amount invested in inventory. The all-new design — which became the H and B series lift trucks — used channel-mast columns in place of tubular columns and used heavier materials and forms throughout. On the B model, customers could get a telescopic mast for lifting objects up to 130 inches (almost 11 feet). "Using this excellent basic channel concept, Crown went on to consolidate the design of its first truck series," says Harold Stammen. "The best feature was probably the interchangeability of parts."

None of this could have happened, say the men who worked on the redesign, if Jim Sr. had not had the guts to decide to scrap Crown's inventory and start over again with a more viable new product. Had he taken a more cautious approach, or been less confident and supportive of the total redesign, Crown might never have succeeded to the same degree.

Once Crown had its two new series, it was able to go to dealers with a nice-looking catalog showing a full line of equipment. Competitively priced, it was "able to compete truck for truck with Big Joe, American Pulley, Economy Engineering, Lee Engineering, and Presto Lift," says Nieberding. "That put us in the material handling business. Later we developed a W series, for 'walkie' (or walk-behind). The walkie lift was the B series, which already had power lift, plus a little power drive unit so you didn't have to push at all. The walkie was powered for travel as well as for lift, but you did have to walk with it; there was no way of riding it." Crown's first walkie fork lift hit the market in 1960 with the product name, E-Z Material Lift. It was marketed initially to small tool-and-die shops and light industry.

The walk-behind

Crown's first "walkie" fork lift, the E-Z Material Lift, hit the market in 1960. Crown marketed it to small tool-and-die shops and light industry.

In 1956 and 1957 Crown was still designing and developing the lift truck more than it was selling it. Not until 1959, when they had annual sales of about $50,000, did it begin to feel like a business, says Tom Bidwell. "We began then to feel this might be a product that will make it. Before that, there were doubts in many people's minds — 'Should we or should we not get out of this?' — because we weren't doing all that well with it. We started with small manually operated lift trucks and we were successful with those, and over the next thirty-five years we built on that, one model at a time. Every year or two we introduced a new model lift truck; a bigger one, a heavier one, a more expensive one.

It just continued to build little by little, one little success on top of another.

"Now when lift trucks were doing $50,000 in sales," says Tom, "the antenna rotators were doing something like $700,000 in sales, so the lift trucks were a small part of the company. But we knew all we had to do was add models and the sales would increase, and that's what we did. I think between '59 and '69 we probably grew at least 25 percent a year. That $50,000 compounded over ten years, and 25 percent growth got us up to where we soon passed total rotator sales."

Growth spurts

In 1961, when Crown's lift truck business began producing revenues, Jim bought the Excello Building, which became Plant 3. In 1972, Plant 4 was added.

By 1961, it was beginning to look as if Crown had a lift truck business. Certainly it needed more room: Lift truck manufacturing required both space and heavy machinery. So in 1961 Jim bought the old Excello Building, a cement-block building at the west end of town, out past the railroad tracks. It was named not for the company's Excello machines but for the old Excello Company which, after experimentally developing turbine blades for jet aircraft engines, had closed the building. The Excello Building became Plant 3. Crown added on to Plant 3 as it needed the space and could afford to: 20,000 square feet in 1963, another 37,000 square feet in 1966, and an additional 18,000 square feet in 1969, for a total 100,000 square feet of plant space. Right away, in 1961, the firm added a wood frame building, a local garage it moved and attached to Plant 3, to give engineering new quarters.

"Up through the 1970s," says Jim II, "we didn't really use architectural services. In those days, we thought that an architect added cost to a project. We didn't realize that if it's done right, the architect not only brings quality design to a construction project but generally saves more with his design than he costs in fees. I remember Verlin Hirschfeld spending a lot of time over at the factory, supervising day-to-day decisions about where Baumer Construction would put an electrical box or drainage tiles. Those are the kinds of things companies do when they are smaller."

At the time Jim Sr. bought the Excello Building he thought the firm had all the room it needed for expansion, but by 1972 Crown had also outgrown Plant 3, and Jim told Bob Marshall to get Plant 4 built and find the machinery for it. By then, the lift

Starting over

Crown's entire line of trucks is descended from Tom Bidwell's basic channel design, which Crown introduced in 1959, when Crown stopped building one-of-a-kind trucks and developed its line of rugged standard models with interchangeable parts. Not only did Crown's redesign make its trucks more rugged, but the interchangeability of parts made production simpler and reduced the investment in inventory. The all-new design used heavier materials and forms, and replaced ***tubular columns*** with a ***channel-mast design***. On the prototype lift table, mast rollers — similar in shape to spools of sewing thread — rolled against a tubular column. But two shapes like that didn't roll well against each other so there was scuffing. With the 1959 redesign, Tom Bidwell (assisted by Harold Stammen) put the mast rollers inside a channel shaped column, which gave the rollers more surface to roll against. The channels allowed a stronger or stiffer cross-section, gave the mast rollers better guidance, and distributed the contact pressures more evenly on the rolling surfaces.

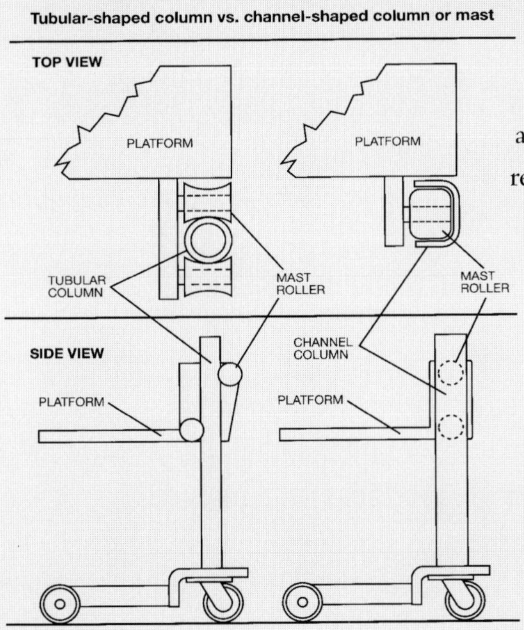

Tubular-shaped column vs. channel-shaped column or mast

The heart and soul behind the development of the lift truck, ◀ ***Tom Bidwell*** was a frustrated artist who turned down an art scholarship in favor of a manufacturing career. Some of his paintings hang in Crown's international headquarters.

Designer Dave Smith called ◀ ***Harold Stammen*** the "brightest, most creative engineer you'd ever want to know. And completely self-taught." Stammen and Bidwell enjoyed "outdesigning the designers."

trucks were beginning to pay for themselves, and Crown could afford to spend $60,000 and $70,000 on pieces of equipment needed for production.

In 1963, however, many still considered Crown's future shaky. Roger Bornhorst, now part of senior management, remembers that when he graduated from high school that year, the school counselor told him that Crown needed someone in Accounting, but added, "I think they are going bankrupt."

"That didn't mean anything to me," says Roger, in explaining why he took the job anyway. "I didn't know what bankruptcy was." Roger started in Accounts Payable and worked his way up. Like many Crown employees, he grew with the company.

How did they get sales going? "We spent advertising dollars," says Tom Bidwell. "We were building our distribution. That was a big part of it. That's when I first hopped on the little single-engine plane and flew all over the U.S., calling on dealers and trying to get them to handle our product. I could get from Erie, Pennsylvania, to Cleveland, Ohio, in one day, where if I was driving I would be exhausted. Of course, in the early days if you had $10,000 or $20,000 or $30,000 in sales it was really something — at least, it was more than nothing. It was a totally different market. It wasn't saturated but there were a lot of competitors, and we were just coming into the picture. Somewhere along the line you started to sort out the ones that were doing a little better and to some extent we didn't know we were doing better than some of the others."

Mostly it was just hard work and attention to detail. "We always had good engineering," says Bidwell. On the first models, Crown stubbed its toe on electrical problems; neither the switches nor the controls were quite heavy-duty enough for the heavy use people make of forklifts. Crown's early switches were designed for actuating maybe six times a day; when people actuated them thirty times a day, their life expectancy was not what it should have been. Crown learned that it needed to go to a heavier-duty truck and it went around retrofitting about 140 trucks in the field with heavier-duty electronic and electrical controls. That took time, but Crown knew it was essential to regain the customers' confidence — to get people saying, "Well, Crown had some problems, but they overcame them." The salesmen knew that it's not that you have a problem or don't have a problem that matters to customers; it's how you take care of your problem. Once Crown handled the retrofitting, it was off and running.

"Our image in those early days was generally that we built a decent piece of equipment but definitely for light-duty work, and that was true," says Nieberding. "That's how it was designed. We were trying to fill a need for the smaller shops that couldn't afford to buy a normal forklift truck. Our lift truck sold for, say, $1,500 or $2,000 and it would do a nice job, where a heavier-duty forklift would have cost $6,000 or $8,000. Crown's was a piece of equipment a small company could afford; if it wasn't available at that low price they would do without." Crown's biggest problem was always that it manufactured basically light-duty equipment that its customers tended to use beyond capacity. If a truck was rated at 1500-pound capacity for load carrying, for example, it might be designed to lift 2,000 pounds occasionally;

but customers tended to use it beyond its normal capacity. Sometimes their requirements simply grew. At one point they had a 2,000-pound load coming through twice a week, and the next year it came through twice a day or even twice an hour. The next thing they knew, their lift truck was struggling because it was being used beyond the intended capacity.

Part of Crown's early marketing strategy was to go to every material handling show it could, with an eye to interesting distributors in handling Crown products. But the marketing budget was slim. They traveled to their first material handling show by train, sleeping coming and going so they wouldn't have to spend money on a hotel room.

At the early trade shows, it was always a struggle for Crown to look as if it were a competitor. Clark was a big force in the material handling industry in those days, and at meetings for its dealers and salesmen Clark passed out "free lunch" passes. It also provided bus transportation between the show and the hotel where Clark was headquartered. There was no way Crown could compete with that, but as soon as the firm had some money for marketing, it tried to do things in a noticeable way. Maybe Crown's salesmen couldn't give out free lunches to everyone every day, but they tried to take their very best customers to dinner. They couldn't put a big group of people up at a hotel, but they would try to put a small group of people up at a better hotel. They didn't have busses to shuttle people back and forth, but for the people they had staying in a better hotel, they provided a Cadillac and a driver. Occasionally the Clark people could be heard mumbling, "Those Crown people know how to do things first class."

Crown started out with a small booth and display, but as it grew in market share its booths got bigger and its displays more elaborate. By the late 1960s and early 1970s, Crown's booth actually dominated the shows. "We had very exotic shows," recalls Eilleen Dicke, "very well-done, not sleazy. We would have dancers perform. We would get a lot of attention when the show would come on every forty-five minutes or whatever. It was a very expensive thing to do, but the other people would just have a little booth, and a couple of men trying to stand there and show their truck. It looked pretty bleak next to this razzle dazzle."

The attention-getting activities drew people to the booth, and Crown got enough of a reputation for its display that people began wondering, "What's Crown going to do this year?" More important, however, "I think we had a better calibre of person at the show," says Jim, "and our people *worked* the booth. Instead of sitting in a corner reading a newspaper, waiting for the customer, our boys were out there on the floor. If somebody came by, our guys would talk to them."

It goes without saying that Crown tried to manufacture a better product than the competition, but what gave Crown a competitive edge was that Crown let dealers handle the product exclusively. Dealers would rather handle an exclusive line than one that anybody can sell, because then they don't have to compete solely on price. ❍

The Long View–and No Committees

According to a study described on National Public Radio in early 1966, firms with longevity tend to be cost-conscious and to diversify only after thoughtful deliberation. They also value loyalty to employees, service to customers, integrity (the path to quality), and home-grown management (executives who come up through the ranks and stay with the firm). That's an excellent description of Crown. The firm has strong central direction but encourages individual autonomy, focused on solving problems and producing a quality product and good customer service. Harold Stammen attributes much of Crown's success to "luck, perseverance, good management, and dealing with what you have at hand. You work to be successful in the scope of where you are at the time. When Crown recognized that it was in the lift truck business, I'd be surprised if anyone envisioned that Crown would be a major lift truck company some day; I don't think any of us expected that. It

Teamwork

Crown's design team worked long and hard on the counterbalanced rider truck (FC series), which won three industry awards (see page 126-127).

was fortunate that Crown stayed a privately held company. I think that contributed tremendously to our success, because we can make big decisions very quickly by just talking to a few people. There are not a lot of stockholders. Large groups of stockholders tend to be reactionary and critical, and want an immediate greater return on investment. There's always pressure to pay out good dividends, which leads to decisions that might be good in the short range but aren't good in the long range — for example, cutting back on capital investments because they need to show a profit this year."

One thing that keeps people at Crown over the long haul is the fact that Crown operates with the long haul in mind. "When the management of a company makes U turns all the time, things will usually slide downhill," says Tony Van der Straaten, a long-time member of Crown's sales force. "When the managers follow a certain pattern, that style will work itself down the ladder. It's an advantage when everyone in the company is pulling the rope in the same direction, and it's interesting how many companies break that rule. Crown has a constant management style with a long-term philosophy — there is a high level of consistency: steady growth combined with a long-term development strategy. *Steady* growth is important. Blowing a balloon beyond its acceptable capacity can create an instant problem: The air enters a balloon slowly, but it can exit quickly."

Steadiness is also valued in the emotional atmosphere at Crown. Tom Bidwell remembers being told in the 1960s, "'Conflicts are not allowed at Crown. Only disagreements.' We sometimes had disagreements but they were ironed out very quickly and simply in one meeting and we went on." Bidwell attributes that pattern to the presence early on of both Jim Dicke and Warren Webster. "It's a style of management that's passed on from one manager to the next. It's

"We were always trying to give overall direction. The best way to get somebody really involved is to have them really buy into it and feel it is as much their idea as yours. We spent a lot of time working on that one-to-one basis. That's changed, of course, now; there are too many people. We substitute for that now with meetings and training courses, but it's not quite the same thing."

—Tom Bidwell

amazing, really, how in thirty years it has just been passed along without thinking about it. It's a team spirit, a desire to get the job done as easily and quickly as possible. It's an informal style of management, a very close-knit team. The decision path around here at Crown is very short. It's right here in this room. We write very few memos in this company.

"I'm struggling tremendously with this in our European operations now because it's typically German and European to create a document for everything. It's also true of a lot of American companies. We're not a typical American company. One of the problems you have with a formal structure, with memos and documentation, is the egos. It seems that once people commit something to paper, it's then locked in. There's some great loss of face if somebody changes their mind. We change our mind all the time here. We say, 'Whoops, that didn't work, let's go in a different direction.' Or somebody may come in and say, 'I've been thinking about it and I don't think we should do what we decided to do yesterday, and we'll say, 'Okay, let's talk about it.' It's just a way of life here.

"We have a problem once in a while when someone new joins Crown from a larger company. We had some of it when we started up our new design center and brought in designers from all around the country. We were getting three- and four-page memos inviting twelve people to a meeting. We'd go to a meeting and even have an agenda. That's amazing, because we never had

"There's no loss of face with Crown anywhere. If somebody says, 'I guessed wrong; the forecasts are wrong and maybe we should change them,' we'll say, 'Fine. Let's change them.' That's all there is to it. When you commit ideas to paper, for some reason they get cast in stone. We try not to do it."

—Tom Bidwell

◀ *Crown Controls Corporation, December, 1962—Seated, left to right: Bob Nieberding, Wayne Brockman, Tom Bidwell, Harold Stammen, James Dicke II, Bob Homan, Paul Dicke, Al Ender (U.S.P.O.), Harry Gilbert, James F. Dicke, Ken Griewe, Adrian Moeller, Harold Opperman, John Koeper, Wilbert Will, Ed Judt, Bob Marshall, John McNulty (G.E.), Jim Uetrecht. Standing, left to right: Verlin Hirschfeld, Jack Lawler, Dick Dammeyer, Duane Hegemier, Hugh Bertsch, Hal Opperman, Arnold Heitkamp, Ken Quinter, Al Schwieterman.*

"We never did hang somebody for making a mistake. Yes, if you make too many mistakes continually they'll try to find you a different kind of job, and after you've done that maybe three times they might say, 'You ought to work someplace else.' But people aren't crucified for mistakes here."

—Tom Bidwell

agendas; we just sat down and talked. It's just a different company culture." Tom would tell anyone heavy into memos, "You know, we may not have enough documentation, we might need a little more, but I don't think we want to overdo it at this company. We're not typical. First of all, we're privately owned. The board of directors practically meets daily here. We don't have a lot of shareholders, we don't have a lot of reports. Decisions are made by two, three, or four people. There are only about five or six people here you have to work very closely with and you don't have to worry too much about communications, so don't try to overcommunicate."

Jim Sr. had a good sense of humor and occasionally interrupted work with a joke. One day in the early years of lift truck development, for example, he called twenty people in supervision, production, and engineering into his office and told them he was holding this meeting to improve the way decisions were to be made at Crown. After the members of the group offered a few suggestions, Jim said, 'I think I have the solution.' He brought out a plastic device filled with fluid, with a little floating ball on top, with a V on top. On one side of the V it said 'Yes,' on the other, 'No.'

"What we do," he said, "is shake this thing up, set it down, and see whether the bubble floats to the yes side or the no side."

When people from outside the firm talk to members of Crown's management team, they are struck by how often the managers say that in all the decades they have worked at Crown, they have never been criticized, and that for the most part they are left to work on their own, with little direction. The Dickes seemed to feel that once you hire somebody to be in charge of something, they're in charge of it, and the only time they'll hear differently is if something doesn't work out or if there's a problem.

"In 37 years I've never been criticized for anything I did at Crown," says Arnie. "I'm sure over the years I've done a lot of different things here but it's a kind of philosophy: You do it and you're not called down. I remember being called together because something Crown was producing had a lot of bad parts in it, so we had a meeting to discuss how the problem was going to be corrected. It wasn't a hell-raising like it might be in some companies. What we did was sit down and try to problem-solve."

"I suppose that's partly true," says Tom Bidwell. "We never did hang somebody for making a mistake. Yes, if you make too many mistakes continually they'll try to find you a different kind of job, and after you've done that maybe three times they might say, 'You ought to work someplace else.' But people aren't crucified for mistakes here.

"We were always trying to give some overall direction, though," says Bidwell. "The best way to get somebody really involved and get the job accomplished is to have them really buy into it and feel that it is as much their idea as it is yours. We spent a lot of time working on that one-to-one basis. That's changed, of course, now. That's almost impossible to do; there are too many people. With 2000 people just here in New Bremen, you lose some of that. We substitute for that now with meeting and training courses that are excellent, but it's not quite the same thing. It's not as personalized."

"It's not that we're not capable of giving people a lot of direction," agrees Jim II. "I'm capable of giving people a lot of direction when I think they're headed in the wrong direction. I try to do it in a kind way." ❍

Appearance counts *Design was what gave Crown a competitive edge, what encouraged and facilitated its growth in the early days. "Most companies in those days didn't believe that industrial products should have a design element," said designer Dave Smith. Above are Crown's PE3000 Series pallet trucks, which in 1994 won the Industrial Designers Society of America's gold award for excellence.*

What's a Pretty Truck Like You Doing *in a* Place Like This?

(1962–96)

9.

"Design became a big factor with us in about 1962 or 1964," says Tom Bidwell. "The lift trucks at that time were basically put together with channel iron, I-beams, and a lot of nuts and bolts, and nobody paid any particular attention to the aesthetics of it. Almost anybody could build a lift truck. You could go to a steel yard, buy a standard I-beam, drill some holes in it, and build yourself a lift truck. Being an unknown company, we needed some sort of edge on our competition, and we decided it should be in the product. I felt that putting some effort into design would give us that edge. We put a lot of emphasis on industrial design; that has worked very well for us. We placed great importance on the appearance of a product because people can't see the interior but they make assumptions about it. If you're buying a car, how do you judge the engine? You look at appearance, the fit of the doors, how well things are finished, and how much attention is placed on detail. If everything that you can see and feel looks good you assume that the manufacturer designed the transmission, the engine, and the other working parts the same way. The image of the purchasing agent with the big cigar saying 'All I want is a good price' is false. If you put six of anything side by side — six clocks, six cars, or six lift trucks — most customers will basically pick the one that appeals to their feelings or senses, assuming price and quality are equal."

That was Bidwell's thinking, and Jim Dicke agreed with him. Crown's lift trucks looked much better than the big lift trucks Clark and Yale were building in the 1960s, and you could see the difference from fifty feet away. Crown felt that attention to detail, in addition to improving the trucks' quality, price, and performance, would give buyers confidence.

Design was what gave Crown a competitive edge — what encouraged and facilitated its growth in the early days. Industrial design was important to Crown even when the firm had no budget for it. It made the most of its slim budget by putting its trust (as it so often had) in talented people who were just launching their careers and were not yet in a position to charge much for their services.

In the early 1960s, Tom Bidwell was looking for something to distinguish Crown's product from its competitors. The only way he could see to distinguish Crown was to package the technology differently. One day in 1960, Deane Richardson, a recent graduate of the Pratt Institute, stopped to see Bidwell, who was then Crown's director of engineering and manufacturing. "Deane was out trying to drum up some business for his struggling new design firm and wanted to do a project for us," says Bidwell. Richardson, together with David B. Smith, another recent graduate of Pratt, had set up business in Columbus, Ohio, as RichardsonSmith. "Why don't you come see us?" said Deane.

Jim Dicke, Tom Bidwell, and Verlin Hirschfeld called on the designers, expecting to find them in a nice office building. Instead they found them in a building that resembled a stable or a chicken coop on Columbus's west end, the inside unpainted but whitewashed and clean. Deane had said, "Since you're going to come and talk about product design, just for the heck of it why don't you bring along every piece of literature you have — your letterheads, envelopes, invoices, and purchase orders." Jim and Verlin wondered why they wanted to see their paperwork. Crown already had a purchase order form; it didn't need another one. But the two designers tacked all the papers up on the whitewashed wall, leaned back, and said, "Look. Every one of them is different. There isn't one thing uniform in that whole mess." RichardsonSmith (better known as RS) developed a corporate identity program for Crown, including a Crown logo and a color system. In the course of doing so, they convinced Crown they could design a lift truck.

"We can't afford you guys. You're right out of college and

pretty expensive," said Verlin.

"Hey, we've got no customers and we need a customer," said Deane. "We think we know how to do this but we have to have a customer."

"I was an easy sell," says Tom Bidwell. "We had a new forklift in mind, but I had a limited budget. We couldn't afford a fancy design firm and they needed a job. Fortunately, Deane agreed to do it." For the modest fee of $750, RichardsonSmith agreed to design a medium duty, hand-controlled pallet truck.

"Bidwell took a flyer on us," Smith told a reporter from *Design News*. "Not only were we still wet behind the ears, but most companies in those days didn't believe that industrial products should have a design element."[1]

Crown's powered pallet truck (which used the drive unit from the M truck) went on the market in 1962. In 1965, it won a design excellence award from the American Iron & Steel Institute. Crown had to squeeze the budget for the travel money to just go accept the award, but Bidwell never had to sell design again. "It became our strategy," says Bidwell. "Dave Smith was so closely involved with our product development efforts that he was more like a Crown employee than a consultant."

Powered pallet truck

This truck, on the market in 1962, won a design excellence award from the American Iron & Steel Institute in 1965. In traveling to accept the award, Tom Bidwell slept on the train rather than pay a hotel bill, but he never had to sell design again.

The pallet truck design by Deane Richardson gained rapid market acceptance and laid the foundation for Crown's new brand image. For many years Crown hired RS to do the industrial design work on every new product development project, in conjunction with Crown's design team. RS focused on human factors and the truck's appearance; Crown focused on everything else. Crown became the first lift truck firm to use a proprietary color scheme, making Crown trucks uniquely Crown. Dave Tompkins of RS, who headed the Crown account from 1963 until he left RS in 1977 to start his own firm, told interviewers for the Design Management Institute case study co-sponsored by the Harvard Business School, "The thing that was so wonderful about Crown is that they cared passionately about how these things looked. The engineers would call us up if they had to make the slightest change. The difference between them and many of our other clients was like night and day." It was a true partnership between design and engineering.

At the time, most people in industry believed that buyers didn't care about design in industrial products. "Why not?" Bidwell asked. "People have design preferences when they buy things for their houses. Why should they leave them at the door when they go to work? I've always believed that it's important for things to look good regardless of what they are or what they're used for. Over the years, though, we've gotten more than good looks for our money. Good design actually lowers your manufacturing costs, and we can prove it."[2]

Crown had started out with the smaller lift trucks partly because it was a small company with limited funds to invest. Its path of product development was then shaped largely by distribution concerns. Crown's distribution network included material handling distributors that sold larger product lines (usually larger, gas-powered lift trucks) and sold Crown's smaller, electric trucks as a supplemental line. Crown's first trucks were smaller than the "walkie trucks" made by the major lift truck producers, Clark, Yale, and Lewis Shepherd.

In 1962 Crown introduced its intermediate-duty counterbalanced WC-series trucks. It got its largest single order for walkie trucks in 1966 when Sears Roebuck Co. of Seattle ordered thirteen units. (In 1969, when Crown began to sell the heavy-duty WC in Europe, it changed the name to WB to avoid jokes about water closets). By 1972, it was making more than a hundred different models of lift trucks, which it sold in 80 countries. Most of its manufacturing was done in Ohio but by 1972 it also had plants in Australia and Ireland.

The walkie

With an order for thirteen trucks, Sears Roebuck Co. of Seattle was the first big customer for this intermediate-duty walk-behind truck, the Walkie Counterbalanaced, or WC, series (changed to WB shortly after Crown began to sell a heavy-duty version in Europe, to avoid jokes about water closets).

In the beginning, Crown's niche was smaller trucks that looked nice and won design awards. The other forklift truck companies, which built larger trucks, ignored Crown, referring to it as the "pretty truck company," according to the Design Management Institute case study. Crown's strategy during this period was to enter market segments where the main design left something to be desired — where there was a market niche waiting to be filled. Once again, its strategy was to sneak into a market without the competition noticing.

"At first," says Harold Stammen, "the major lift companies didn't complain that their distributors were also selling Crown trucks. As Crown began to grow and wanted to expand its product line, we started to produce trucks that were competitive to what the dealers were already handling. We introduced bigger trucks that were directly comparable, for example, to a heavy-duty Clark walkie truck. That didn't affect the big competitive manufacturers too much because the gas trucks and electric rider trucks were their bread and butter."

Crown decided to move forward by developing a counterbalanced rider truck, a heavy-duty lift truck the operator stood on to drive. This truck took a noticeable jump in size as a Crown product because it had a lift capacity of 3500 pounds and it had to accommodate a rider-operator. A counterbalanced rider truck has a mass of iron opposite the load-carrying end to keep it stable while lifting. It's a more complex and expensive product than the standard walkie trucks. Crown selected the end-control counterbalanced rider truck for strategic reasons. This was a niche in which the major lift truck manufacturers did not have a strong product, so Crown could sneak into the heavy-duty rider part of the market.

"I remember starting on the design of the vehicle as a rider reach truck (a stand-up truck with extendible arms) and changing course after we were maybe three or four months into product development," says Harold Stammen. "We changed course to a stand-up rider end-control counterbalanced truck because Tom Bidwell began to feel that if we introduced the reach truck the major lift truck companies would object to their dealers handling Crown and the dealers would be forced to make a decision as to which company they would keep in the dealership. We didn't think we were strong enough yet that the dealers would decide in our favor, so we switched our design focus to another product — an end-control rider counterbalanced truck. In this area, the major lift truck companies had a weak product; it was not a large market segment to them so it was less threatening. We didn't want to upset the distribution network that we were part of; we wanted still to be considered a supplemental line. We kept a low profile so as not to upset the current distribution arrangement.

"As we completed the end-control rider counterbalanced truck (the RC series) the next step was to go into the three-wheel sit-down rider counterbalance (the SC series), which used the same new drive train but with a different arrangement of components. We expected, during the design of the stand-up truck, to be able to use the same components on the sit-down. This was a natural evolution, but gradually we became more threatening to the major manufacturers. At some point, as we got into larger trucks the major companies began to object to their dealers selling Crown's big trucks," says Stammen. "At some point the dealers did have to make a choice. Some dealers made a choice not to give up either line. They just set up another dealership and owned both."

"It was a major thing for us when Crown came out with the first lift truck that had a seat on it," says Tom Bidwell. "We were known then as a walkie/stacker company, and all of a sudden Crown put a seat on the lift and you drove it around. It was a little frightening at the time. We wondered if we were doing the right thing because our competition was much bigger than we were."

Crown designed the new models in that sequence partly to avoid tipping its hand about what it had planned. The RC truck — the one you stand on and drive in side-stance position

— was done before the sit-down counterbalance or the reach truck because it represented a smaller share of the market, one that nobody would get too "shook up" about. The drive mechanism and dynamics — motors, gears, transmissions, and electronics — of the two counterbalance trucks were the same.

"As with all new products," Tompkins recalled for the Design Management Institute case study, "Crown began with analyzing everyone else and coming up with what the new truck would be. It was like a dot-to-dot drawing, this information (like head length, turning radius, lift capacity) being the dots." Tompkins and his RS colleagues did even more research. "At this time," says Tompkins, "every other [stand-up rider] truck had this bizarre condition that, when traveling in reverse, the operator had to turn around and operate the controls from behind his back. We tossed around all sorts of ideas about what to do about that problem. Then we said, 'In theory, you could stand sideways and see both forwards and backwards just by moving your head. But then how would you operate the controls?'"

The side-stance design

An industry first, the award-winning RC truck was an ingenious solution to an old problem, that the operator traveling in reverse had to turn around and operate the controls from behind his back. Standing sideways, the driver could see both forward and backward by turning his head.

The side-stance design — an industry first — was an ingenious solution to the problem. On Crown's Rider Truck, which the RS design team designed in the early 1970s, the driver stood sideways for easy front-and-back vision, and steering was controlled by the left hand.

"Tompkins' team then built a full-scale mock-up and had people of different sizes interact with it," said the Design Management Institute case study. "They ran a vision plot to see where blind spots were, and they tested various configurations of a multifunction control system. To accommodate nearly a dozen functions, they developed two controls. The main control lever for the right hand moved the forklift forward and backward, up and down. It had two buttons — one for the horn and the other for the tilt mechanism. Steering was controlled by the left hand. When Tompkins offered this radical proposal to Crown, 'it took a while to convince them.' The new RC (Rider Counterbalance) truck, bombarded with criticism by the competition, offered customers their first 'real choice' in lift trucks." Within four years, the truck captured a strong market share.

After years of slow development, by 1974 Crown's lift truck business finally began to take off, and the firm began expanding at a rapid pace. On the street where Plants 3 and 4 were, Crown bought a big open field plus six or seven more houses. Crown spent a lot of time convincing people to sell Crown their houses so Crown could expand. "Where Plant 4 and 5 are, we thought we had plenty of room because we were never going to get that big," says Verlin Hirschfeld. "Now we were being forced either to find a new site in the country and start all over again (which we didn't have the money to do) or to convince people to sell their houses to us." Crown would let people remain in their homes for the rest of their lives but might build a plant eight feet behind their house. As Crown grew, it numbered new buildings in order of expansion. In 1996, construction was beginning on Plant 8.

After the RC truck, Crown's next major success was the rider reach (RR) truck, introduced in 1980. Starting with the RC and SC trucks (introduced in 1972 and 1974) had given Crown valuable manufacturing and field experience, which made designing the reach truck easier. The RR was designed for the narrow aisles of warehouses and distribution centers. Unlike the counterbalanced truck, which has a mass of iron opposite the load to keep it stable while it's lifting a heavy load, the rider reach truck has legs, or outriggers, in the front to keep it balanced. The rider reach truck became a Crown showpiece, winning a design award from the Industrial Designers Society of America in 1980 and, in 1990, the "Design of the Decade" award from the Industrial Designers Society of America. Crown builds a lot of rider reach trucks.

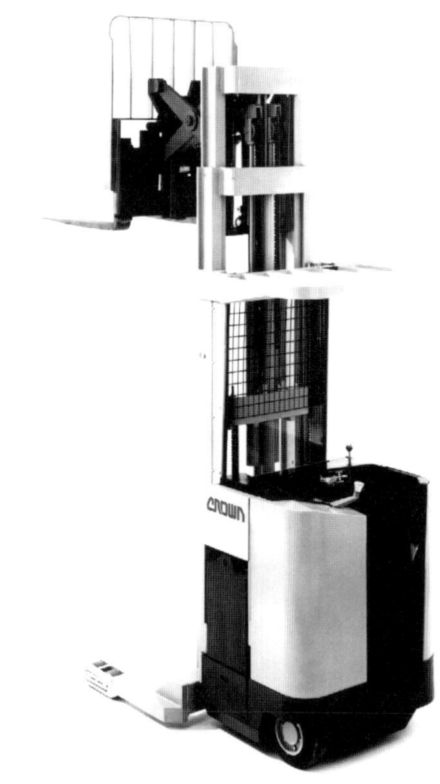

The reach truck

Introduced in 1980, the rider reach truck was designed for the narrow aisles of warehouses and distribution centers. Legs, or outriggers, kept the truck balanced. A Crown showpiece, the RR truck won the "Design of the Decade" award from the Industrial Designers Society of America.

The reach truck looks a lot like the RC truck except that the RC truck has a fixed fork and outrigger, and the reach truck has a panagraph, a scissors-type arm that reaches out with the forks to pick up a load. The panagraph collapses and retracts with the load. You can pick up a pallet, retract the panagraph, or push it out without moving the lift truck. The RC, because of its need to be counterweighted, is longer and takes up more room. It is used more for loading and unloading highway trucks at the docks. The RR is used primarily for working in narrow-aisle areas.

The RR series allowed Crown to move up from its position as a niche player and become a true competitor with a full product line. And with the introduction of narrow-aisle reach truck models, Crown became a leader in this market segment.

Crown employees threw themselves into projects like the RR series. Harold Stammen tells a story explaining why they did so. In the early 1960s, Harold was developing a drive train for a combination tow tractor/personnel vehicle (and eventually for use on the narrow-aisle reach truck). "In testing the drive train," Harold recalls, "we discovered that the efficiency of the drive unit was poor. I spent a lot of time designing it, having the prototype built, and testing it to determine what was causing the inefficiencies. We had put a lot of money into it and in the end I concluded that there was a serious flaw in the design. This was going to be a major setback for Crown and I felt responsible for it. I felt sure they were not going to take it lightly but in my mind it was necessary to report it, so I went into Tom Bidwell's office. I had no more started to explain when Jim Sr. walked in. I went through my presentation as to how we needed to provide a quality truck, and this drive unit would not give the performance wanted and should not be used for production. I explained why the design was bad, why we couldn't salvage it, and why we had to start all over, with a complete redesign, from the ground up. At the end there was silence and I wondered if this would be my last day at Crown.

"There was some discussion and then Jim Dicke thanked me for my forthrightness. 'You know,' he said, 'you could have proceeded with the design as if there was nothing wrong with it. That would have put Crown at considerable risk later on. We would have had products out in the field that would have created unhappy customers.' He thanked me for my frankness. They were in agreement about scrapping the design and starting over. It gives you a really good feeling about the company."

Jim had a way of making people feel listened to and respected, but his strengths went deeper than that. They didn't feel they would be hung out to dry if they made a mistake. "Jim knew how to bring out the best in people," says Harold. "He used their talents, and working at Crown you felt that you were in it together, that everyone was together as a team. He did a lot to bring that about. He had a strong dedication to the company. You would do the absolute most you could to respond. I didn't see Crown as a place where I worked from eight to five. I *lived* Crown. Many other people did the same thing."

"Crown's approach to industrial design stood in stark contrast to that of most other industrial companies," reports the Design Management Institute case study, "who typically hire design consultants to style a product after all the internal engineering was completed. Often design consultants had little say in defining the product concept and were seldom in a position to influence the engineering. With Crown, the situation was quite different. They would call in the RS team as soon as they had a clear idea of a new product target. Both RS's industrial designers and Crown's engineers would assess the user's needs and the competitors' products, from different points of view. Then RS would develop a proposal and work with Crown engineers during product development."

The case study quotes Don Luebrecht, then a senior project engineer, about the secret of Crown's design team: "We work together very well. They always listen carefully to us and we always listen carefully to them. Sometimes we have to re-do some of our engineering to accommodate their designs; sometimes they have to change their design recommendations to accommodate our engineering. Sure, there are minor conflicts, but we always manage to work them out. There is always a lot of give and take. Of course, they always go home at the end of the day."

In a sense, Crown now takes industrial design almost for granted. For more than thirty years, it designed trucks in conjunction with RS. Now it has an in-house design center working fulltime on design. It won't settle for poor design, but it puts increasing emphasis on engineering and new technologies to stay ahead of everybody — not just domestic competitors but foreign competitors, particularly the Europeans.

"Now of course, the whole industry is paying attention to design," says Bidwell. "Now design alone cannot carry Crown through because other companies can also hire good designers. Now Crown is putting far more emphasis on what they call 'under-the-hood' items: better motors, better bearings, things that will make a quality product last a long time with minimum service. Crown is putting a lot of emphasis on that, in addition to having a good product. It is sort of like the German mystique: If it's German-made — Mercedes, BMW — it's good. Crown is going for that same effect. They have an attractive product but appearance is not the only thing."

"We've always had good designs," says Bob Nieberding. "We learned from some of the earlier mistakes. Even when we had full-fledged development, when we did mock-ups and allowed four years to engineer a product and do all of the testing and evaluating, eventually when it hit the market there still might have been a hickey here and there — but we weeded out 90 percent of them in the process of development. Once we got big enough and had enough engineering staff to do that, we generally came out with a really good product, right from the word 'Go.'"

Industrial design became part of Crown's developmental strategy whatever it did: product engineering, corporate literature, or advertising. When Crown first hired an advertising agency, it had a small budget but insisted on the agency's creative best — no compromise. Crown used its strength in design and styling to develop its market, and it developed a strong marketing organization. Crown puts a lot of emphasis on corporate communications with its employees, its suppliers, and its customers — all of whom, it feels, have something useful to say about how to maintain high performance and customer satisfaction. As competition intensifies, the employees, suppliers, and customers help Crown stay ahead of the game.

Crown's branch operations proved to be an excellent training ground for Crown executives dealing with the retail side of the business. The branch network also allowed Crown to develop a closer relationship with its customers and to keep its finger on the pulse of the end users. This focus on customer needs was paramount with the firm. When one of Crown's major customers met with Crown in New Bremen, for example, twelve of its vice presidents came in with no other purpose than to see Crown, meet its management, learn more about the company, and let Crown know more about their operations, so that Crown would have their needs in mind when they were designing new lift trucks. Crown was already doing good business with this customer but was interested in knowing what their problems were and what Crown could do to help solve them.

"In research," says Tom Bidwell, "for improving our new models for different types of lift trucks, we'll go in and visit a customer's operations, we'll videotape them, we'll talk to the operators, the managers, the warehouse managers, all the way up to the vice presidents. We'll ask them where they think their business is going in the next ten years and what their problems are. We gather this information and see what we can do that fits in with their plans. If we're designing a lift truck today that's going to be used for the next ten or twenty years, we need to know as much as we can about how that application may change in the next twenty-five years. We're doing far more of that today than we ever did before."

Because Crown's design process is user-oriented, the design team extends from sales to engineering. Everybody at Crown is sensitive to customer needs. Both engineers and industrial designers go out into the field to talk with the owners, the operators, and the service personnel; they find out how the hardware and systems work and what the buyers of equipment want the equipment to do. "The design process constantly reminds us why we are in business," says Tom Bidwell, "by putting the focus on the customer and the user. Meeting customers' needs and expectations is paramount. The manufacturing and design team has to think beyond producing the lift truck to understanding how the customer is going to use it."

"There are some constants in the firm that make the team approach work," David Smith told the Design Management Institute. "The first has to do with personal relationships. Over the years management has continued its unwavering commitment to high standards and design excellence. And this has permeated the entire organization. The engineering group has a strong visual orientation and a great sense for design issues. There is little turnover, and the hallmarks of our collaborative relationships are experience, trust and openness. There is equally strong support from sales and service. They enthusiastically endorse the best ideas regardless of where they come from in the team and contribute with a constant flow of information about customers, users and their needs. Development is based on thorough research with extensive input from the manufacturing unit, people who strive to make the product just like it was designed, and from suppliers with whom Crown nurtures close partnerships. The surprise is that all of this just seems to happen, almost informally."

In 1984, Crown introduced the enormous man-up turret sideloader (TS) — one of the most advanced trucks of its kind in the United States. The turret truck works in very narrow aisles, yet can lift pallet loads as heavy as 4000 pounds — taking the operator up with the load —

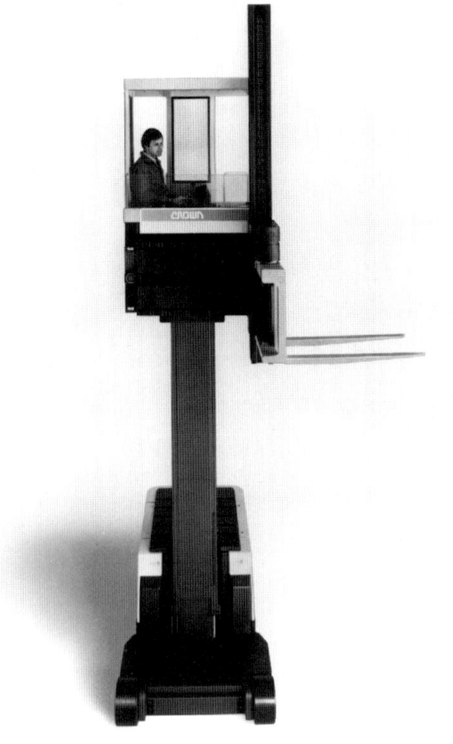

The turret sideloader

Turret-style trucks already existed in Europe, where limited space necessitated high-density storage early on. Many early turret trucks were locked onto a track; Crown spent years in research and development before coming up with this man-up turret sideloader, a free-wheeling truck.

as high as 45 feet (although at maximum elevation the capacity is reduced). Turret-style trucks already existed in Europe, where limited space had forced more concentrated storage earlier than was necessary in the United States. Crown engineers and RS designers had gone over to Europe to study the European turret trucks and other industrial equipment. "The concept of high-density storage started in Europe, where they don't have acres and acres of ground," says Bob Nieberding. "They had denser storage conditions and their only choice was to go up. In Europe, however, at one time the material handling equipment was more or less a trolley that would lock into something on the ceiling and on the floor and would move along a track." The equipment being locked onto that track — like those ladders you sometimes see in a bookstore that move along a track parallel to the bookshelves — was what kept it stable. They were also producing a *free-wheeling*-style truck, a truck that could move out into an open area all by itself and could also place a load up 40 feet high. "Today, for the most part, the U.S.-made products are the free-wheeling type," says Nieberding. They are not locked onto tracks.

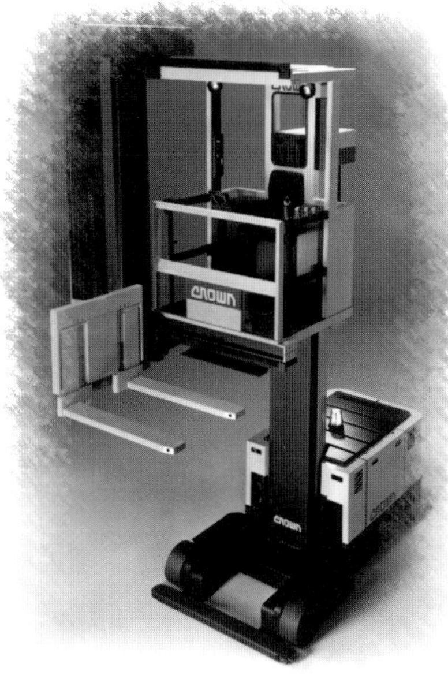

The turret stockpicker

This elegant turret stacker also captured several design awards. It has enough weight in the truck to keep it stable, even with the load coming out off the side.

In the research stage, Crown visited European and U.S. facilities as early as 1972 to study how current products were used and to interview the equipment operators about how they did their work. Their focus in designing the turret sideloader was to make the equipment operator-friendly, so they also conducted ergonomic studies. Intense design work on the turret sideloader began in early 1978. Using all the information they had gathered, the designers made sketches of design features, including the layout of controls. The design and engineering team evaluated every design decision with careful thought to such details as the shape of the control levers and the location of the drive motor and transmission. The team worked several years before arriving at a product they felt was ready to prototype. After an additional four years of design, engineering, and manufacturing work, and after custom designing and manufacturing many of the components, Crown introduced the turret sideloader to the market.

Four years later, it introduced the elegant turret stockpicker (TSP), which soon captured several design awards. The turret stacker has enough weight in the truck to keep it stable even with the load coming out off the side. "At some point," says Nieberding, "you have to wonder if there is a point of diminishing return because you have to put so much weight into a truck in relation to your payload. With a normal lift truck that doesn't lift very high you may have a 5000-pound truck that can lift 4000 pounds but when you get into these trucks that lift high and reach out to the side, a truck may have to weigh 9000 pounds to deliver a payload of only 2000 pounds. So you are handling less weight with more mass and that gets very expensive. The only place that this high-density storage equipment is justified is when the machine costs more than offset the costs of construction and land procurement."

Why did Crown make electric lift trucks instead of gasoline-powered trucks? "In the market we serviced," says Jim Dicke, "we were selling small forklift trucks. It wasn't practical to consider a combustion engine on that small a truck. We were using golf-cart-sized batteries, a small motor, and a little hydraulic pump that could lift 1000 pounds. That was the market we started in. The market once favored gas-fueled lift trucks because electric trucks are more expensive initially. But they are less costly over the life of the truck: They are easier to maintain, they last longer, and they are plugged in overnight to recharge so they don't use expensive fuel. Now the market is split about fifty-fifty. The only time we would even consider the gasoline-truck business would have been with the last truck we came out with, the four-wheel sit-down rider truck that lifts 6000 pounds. That's a big market for us today, and a lot of those trucks are sold but the market is dominated by five to seven companies, about four of them Japanese, so there is not much margin left in it. Everybody is having a hard time in that market."

"Our equipment is all used indoors on concrete, so electric equipment is fine," says Tom Bidwell. "When you get into gas trucks you are generally moving outside of buildings and you are handling different kinds of terrain."

The Levitz stockpicker

Unlike Harold Stammen (who "couldn't relate to an antenna rotator"), general manager Verlin Hirschfeld didn't have much faith in lift trucks, until the day Levitz, the big furniture discounter, placed an order for 67 stockpickers. "That caught Verlin's attention," said Harold.

Jim Sr. always had vision and plenty of ideas; with the lift truck he also got focus. "At one point," he says, "heat regulators were 100 percent of our business. Then they dwindled off and antenna rotators became 90 percent of our business. They slowly dropped off as lift trucks became more and more a part of our business. Contract work amounted to 15 or 20 percent of our business at one time. Today, 95 percent of our business is in material handling. And sales on lift trucks just keep going up."

This is happening, explains Verlin Hirschfeld, because "more people are buying lift trucks. Nobody lifts anything more than 25 pounds today. You can go anywhere — to Sam's Clubs and places like that — and you'll see lift trucks running up and down the aisle. I have a lot of fun in these places: I tell them the paint is wearing off, it's about time to buy a new one. 'No,' they say, 'they're a good piece of equipment.' If I see competition in there I tell them they got the wrong one."

"I don't think Verlin took much interest in lift trucks as a product at first," says Harold Stammen. "His focus was more on the subcontracting, which he had a lot of responsibility for and was very close to, and on the antenna rotators. He also handled general business decisions, financing, and the running of the company. He pretty much left lift trucks up to Tom Bidwell to handle. I never saw much of Verlin until the day in the spring of 1971 when we received our first order for stockpickers from Levitz (the big furniture discounter). That was a big day." Stammen and Crown's sales manager under Tom Bidwell went to Phoenix for almost a week to meet the buyers for Levitz and look at applications and equipment in Levitz stores. Convinced Crown could meet Levitz's needs for a better truck, Stammen worked with the special lift department to come up with a lift truck that would do what they wanted. They got an order for 67 big trucks. "*That* caught Verlin's attention."

Before that, "it wasn't unusual for us to hear that if it weren't for rotators and subcontract work, we wouldn't be in the lift truck business," says Stammen. "'The rest of the company is supporting you fellows.' Then all of a sudden lift trucks started to really take off. Major businesses started buying. We'd begun to have some success before the Levitz order, but Levitz was the thing that really put us over the top and convinced people like Verlin that the lift trucks could make it on their own. For the first ten years the lift trucks were subsidized by other Crown products. After that, they started making money on their own."

Today Crown Equipment is an international company with $700 million in sales. In late 1994, Crown appeared for the first time on Forbes' list of the top 500 private companies in sales. "Everybody looks at us as the overnight sensation," says Tom Bidwell "but it took thirty-eight years to get there."

Crown now produces a full line of lift trucks, ranging from the small hand pallet truck (which sells for about $575) produced in Galway, Ireland, to the rider reach truck (which sells in large volumes, at prices ranging from $25,000 to $30,000) to the huge turret truck (at prices ranging from $60,000 a truck to about $93,000. The company offers several dozen different trucks in ten product categories.

Crown makes 35,000 lift trucks a year. "That's a lot of trucks," says Arnie Heitkamp. "And we have a serious backlog of orders. Sales works hard at that. Of course, some of the backlog is stock orders that dealers want, as opposed to customer orders."

Crown is the largest manufacturer of "walkie" (walk-behind) lift trucks and stockpickers in the United States. It is well known for its innovative narrow-aisle reach trucks.

Crown ranks third in sales of all types of electric lift trucks worldwide (including the United States). It has the largest share of the U.S. market in lift trucks. Crown ranks tenth in the world among all lift truck manufacturers — that includes both gas- and electric-powered trucks — although it produces only electric trucks. Japanese companies are major players in the global market but tend to be stronger in gas and diesel trucks than in Crown's market, the electric trucks. "In the narrow-aisle lift truck field," says Jim II, "we're number one. But throw in the 6000-, 8000-, and 10,000-pound sit-down rider trucks or the gasoline-powered trucks and we drop to sixth place." In sales of high-lift stockpickers and the walkie stackers, Crown's U.S. market share is consistently the leader, and Crown has begun to increase its sales in foreign markets.

Crown employs thousands of people worldwide. In Ohio,

it is Auglaize County's largest employer; hundreds of employees live in New Bremen alone. And Crown has been lucky in economic downturns, says Verlin. "Generally, what we've done is expand, expand, expand. When there's a downturn, we don't expand. We just sit back and cool it. We've had a few bumps and layoffs, but not many, really, not compared with the rest of the country."

Crown is no longer in a small-niche market, sneaking into the industry. It is now a multinational firm producing a full line of electric lift trucks. "The key challenge in this industry is to stay on top," Jim Moran (then sales director and now senior vice president) told the Design Management Institute case study researchers. "If you look at lift trucks historically, you see that no one has ever sustained market leadership over time. The leaders have always been toppled by someone trying a new approach. For years, we had the luxury of being able to sneak up on market leaders who had become lulled by their own success. Now we've become the market leader that others are gunning for."

Inevitably, at a certain point, Crown began to be copied: Other companies came out with lift trucks painted in similar colors and with a copy of Crown's control handle. In an ad campaign in the early 1990s, Crown turned that to an advantage in an ad that focused on the features being copied, with the caption, "Take a close look at Crown. You can tell the difference."

1960

1961

1962

1966

1974

Crown's identity

Design was part of Crown's developmental strategy in product engineering, corporate literature, and advertising. Crown was the first lift truck firm to use a proprietary color scheme, making Crown trucks uniquely Crown. In 1994 Crown earned the first Beacon award for excellence in integrated communications.

And you can. Since that first design award in 1965, Crown has earned more than twenty-five design awards, often for more than one truck. Naturally, this impresses customers. Crown emphasizes good mechanical design, appearance, graphics, and ergonomics. But most important, they manufacture the safest products in the industry. They do that because people want safe products and it is worth the extra money to be sure they are the safest. Crown is especially strong in the area of serviceable design — for example, the way doors swing open, so you don't have to disassemble the truck to get to a particular part to service it. And now Crown trucks have such features as ergonomically suitable seats in which the workers are comfortable, and can turn around and look and drive backwards easily.

"Today," says Harold Stammen, "there is a much closer balance of activity between new product design and redesign. For many years we had to continually prove ourselves. Every time we introduced a new product, we were breaking new ground for Crown. Our attitude was that we could not afford to come out with a mistake. We were not that big a company that we could absorb the risk, so in the process of introducing a new product we looked at everybody else's products, we talked to customers, we had the advantage that we didn't carry any old bad baggage and it was critical that the new product we came out with did all and hopefully more than the customer expected. With an all-new product, there's a greater risk that something might be missed. We had to make certain that didn't happen, and we were successful.

"Now that Crown is into product evolution, changes are more incremental," says Stammen. "It doesn't mean it doesn't have to be right, we just take off smaller bites rather than enormous bites. Change has to come faster. Before, we would take five years to develop a product; now we have to do it in two or three instead."

"Like many companies, Crown spends more on training and re-training today than they would have a decade ago," says Jim II. "Young people entering the work place may have several careers even within one company over the course of employment and will more than likely have to be involved in continuing education programs.

"Times have changed for Crown as for everyone," says Jim II. "The pace of design changes for industrial products has gotten much faster. Once an industrial product — a lathe or a lift truck — might have a ten-year life cycle without significant change. Now the engineering for industrial products changes as quickly as the engineering for consumer products.

"Not everything in material handling has been invented," Jim II continues. "The whole industry continues to grow slowly, but it is going to be a different industry. The field is really exciting. We have an opportunity to strengthen our leadership in this worldwide industry. There are so many more technological innovations to come in all industries. Already today lift truck operators use centralized computer systems to determine a steady flow of production. There is computer selection of materials, computer reorder picking, designs for trucks that take storage up 45 feet. There are lift trucks for large refrigerated warehouses in which a heated cab allows a man to ride around inside all day. It used to be that people would work inside a frozen food warehouse for only part of the hour. There would be three different shifts going in and out because it was so cold people couldn't stand it. Having a lift truck with a heated cab is one way the industry had to change. And look at what Wal-Mart stores have done. They've become the number one retailer in the United States and their needs are much different from those of a little store. The connection between the lift truck operator and the job task will become significantly more sophisticated. The great challenge for American technology remains to make the job interesting and more productive." ❍

[1] *In discussing Crown's lift truck design, I do not always distinguish between industrial design (which focuses on ergonomics and aesthetics to add product and user value) and traditional engineering design (which focuses more on function, reliability, manufacturability, and cost) — partly because Crown's engineers and its industrial designer worked so closely as a team that the traditional distinctions tended to disappear.*

[2] *Quoted in the Design Management Institute case study.*

How does an industrial truck handle the load?

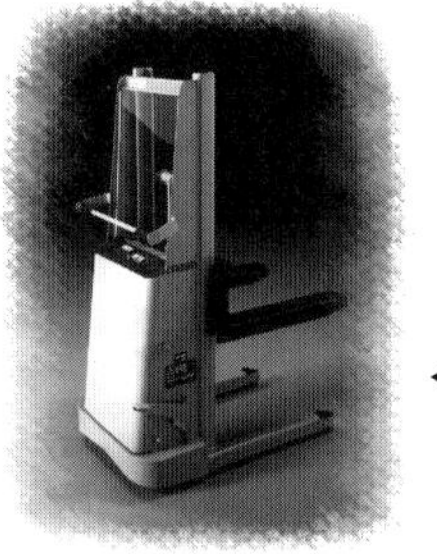

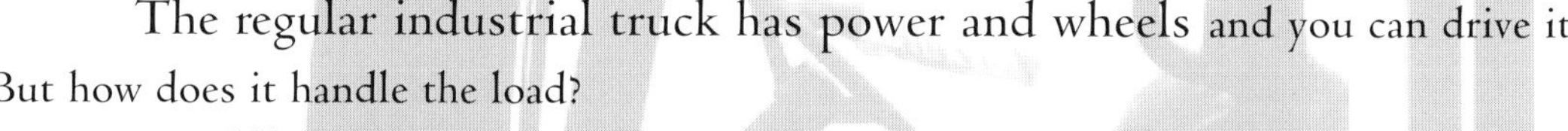

The regular industrial truck has power and wheels and you can drive it. But how does it handle the load?

A ***straddle truck*** has legs, or outriggers, out in front that lengthen its base to give it stability. The straddle truck — a narrow-aisle truck — is generally shorter overall than a counterbalanced truck, but has to straddle the load to pick it up. The truck's drive end isn't very long. With the straddle truck, you can take a load, make a right-angle turn in a narrow aisle (up to eight feet wide) and put the load in the rack. Because the truck has a narrow base, it can get down a narrow aisle.

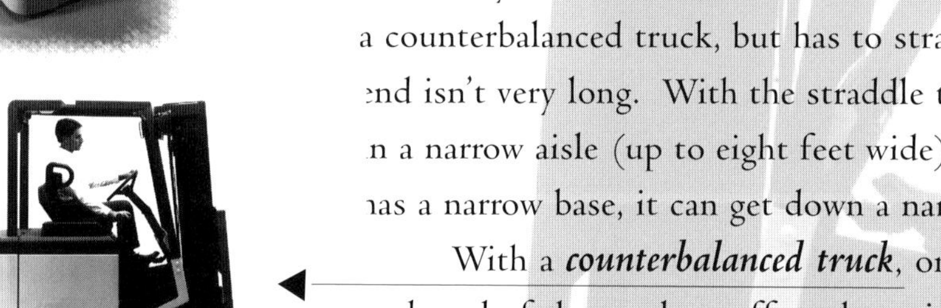

With a ***counterbalanced truck***, on the other hand, you have enough weight in the back end of the truck to offset the weight of the load, so the truck won't tip over. The counterbalanced truck does not have straddle legs, so it can engage the load and storage rack without concern for the size and alignment of the straddle outriggers.

The ***reach truck*** has a fork carriage mounted on a pantograph that extends to permit the pick-up and deposit of loads without interference from the straddle legs; it can also retract the load for transporting and turning in a narrow aisle. Combining benefits of both the straddle and counterbalanced trucks, it is the most popular truck for narrow-aisle handling.

Crown ***stockpickers*** are high-lift orderpicker trucks in which the operation controls are stationed on the truck's platform. That platform (with the operator on it) elevates along with a pallet or platform carried on forks extending from the operator's platform. Typically the aisle is designed to be about a foot wider than the truck. The driver can have steering aided electronically or by guide rails that permit hands-free aisle travel (with the platform elevated up to 30 feet). The operator controls the truck, making stops along the aisle to pick goods from the rack to fill a particular customer's order. When the order is completed, the truck brings the goods to the dock for shipment.

Crown's operator-up ***turret trucks*** are designed for use in very narrow aisles — aisles narrower than reach trucks serve. A turret truck stores and retrieves pallet-size loads weighing up to 4000 pounds in an aisle as narrow as five feet across, and elevates them to heights up to 45 feet. The operator controls the truck from inside the cab, which also elevates (for close visibility and more accurate control of load location). The load-handling device can rotate forks 180 degrees to face either side of the aisle without the truck turning (hence the name "turret truck"). The forks can traverse the aisle width and move into rack storage to pick up or deposit the load. Because there is little clearance between the sides of the truck and the sides of the aisle, the truck is guided electronically or by rails within the pick aisle. Outside of the pick aisle, steering control is returned to the operator. Some trucks are equipped with an enclosed, heated operator's cab for use in freezer warehouses as cold as -20° Fahrenheit.

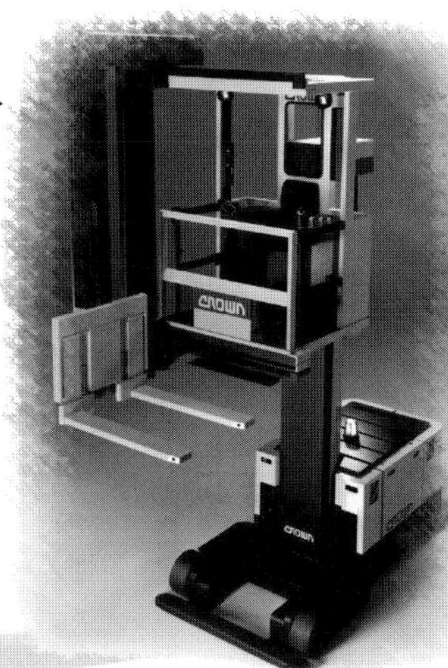

Getting around *In the days before cars and lift trucks, New Bremenites would take their local streetcar to St. Marys to see the doctor or get their teeth done. The town was connected by a spur to the route for longer-distance trains, including the one between Cincinnati and Detroit. Opposite, Monroe Street.*

An Image *of* Quality

10.

Crown in its early decades valued ability more than credentials and in return got not only amazing loyalty from its employees but performance far beyond even the employees' expectations of what they could do. Although Crown has grown enormously in fifty years, it still has its corporate headquarters and major manufacturing facilities in New Bremen, a small farming community with a strong, intelligent, caring group of citizens. Many people see industrialization as replacing beautiful countryside with ugly factories; this hasn't happened in New Bremen, partly because the Dickes are so community-oriented. When Crown had to expand its corporate facilities there was limited office space available downtown, but instead of building new office space Crown chose, for the most part, to restore old downtown buildings, most of which had been built in the 1850s and 1860s, during the Miami-Erie Canal days.

Hot times

Crown has made New Bremen part of its competitive strategy, restoring old buildings downtown with an eye both to authenticity and modern building code. Among the historic sites restored was this old fire house, built in the Miami-Erie Canal days.

Several departments — including accounting, sales, and design — are located in refurbished old buildings downtown. For years advertising was located in a former dry goods store. Personnel is in a former candy store. A house south of Plant 2, which in the last century was the home of a prominent local businessman, was remodeled and expanded and now houses Crown's legal department. Crown was careful in enlarging it to match the bricks for the addition to the old bricks already on the building. Restorations are done with an eye to both authenticity and modern building code. New Bremen became a canal town in the nineteenth century, and in Crown's early days an old canal boat was still rotting away at its moorings. The first wave of immigrants, in the 1830s, were German Protestants who traveled by canal boat to Piqua, Ohio (thirty miles south of New Bremen), then finished their trip by wagon or on foot. By 1843, immigrants had begun reaching New Bremen in canal boats, although most of the German Catholics got off just before New Bremen, in Minster or Fort Loramie. The Dicke family arrived with the second wave of German immigrants, in the 1840s and 1850s. New Bremen remained largely a Protestant town until shortly after World War II, when it got its first thriving Catholic church.

New Bremen's modern identity began in 1945, when the Dickes launched Crown Controls. The name Crown came from a variety of mink that the Dickes had raised on their farm before World War II when the mink's light reddish brown fur was popular. "Crown" would do, they decided, until they thought of a better name.

Downtown New Bremen is only three blocks long from one end to the other. The buildings tend to be one-, two-, and three-story red brick and stone buildings with white woodwork. Some are plain, some have bay windows, and one (the Crown store — formerly the town bank) has white columns. For a town with a population of about 3,000, it's an unusually dignified downtown area. Half of Crown's employees live in New Bremen; the rest live within 40 miles of Crown, in the small and medium-size towns of central Ohio. The county airport, the Neil Armstrong Airport,[1] is located in New Knoxville, six miles away. The nearest major airport is 45 miles away, near Dayton, which is 50 miles southeast of New Bremen.

Restored

Not only was the old fire house restored but New Bremen could boast the most complete restoration of a fire pumper anywhere in the country.

Jim II oversaw the design and building of Crown's international headquarters, a new building erected in 1994, which combines old-world and airy new-world styles. Crown owns about twenty buildings or roughly two-thirds of the downtown area, including the opera house, a hotel, and a canal warehouse. The firm has gone out of its way to preserve an almost residential look. Crown nurtured a new residential housing subdivision (with wide streets and sidewalks) on the east side of town where Jim II and Janet's son Jim III and his wife Katy have built a home. It also launched a shopping center on the south edge of town on land that once was part of the Dicke family farm. In recent decades, on local farmland bought at public auction in Minster, executives of Crown and of other area businesses, knowing the importance to a growing company of providing amenities that will make its employees happy to be living in the community, helped install an eighteen-hole golf course.

Crown has always been seen as a neat and tidy facility, its grounds in good condition. Tom Bidwell calls Jim Sr. Crown's downspout inspector. "Every time he comes through the factory, the only thing he looks at is to see that the roof isn't leaking, that everything is painted and nice and fresh and clean. And that's good because those things really count in an employee relationship. When a worker comes in and sits on a work bench and he's got a three-legged stool to sit on instead of four legs or it's got a crack down the center and it pinches you every time you sit down, those are the things that make you say, 'I'll look for another job.' We always try to see that doesn't happen. Our factory is cleaner than most factories you'll see in our type of business. We always try to see that the canteens are nice, that people have good working conditions, as nice as we can make them. That has always been very important to Crown management."

All of this contributes to an image of quality. The people who live and work with Crown value a quality environment and so do Crown's customers. They see the same quality in the townspeople that they see in Crown products. "To sell our products," says Jim II, "we bring customers here to see the lift trucks manufactured. The fact that New Bremen is an attractive community helps us sell the product. Crown has made New Bremen part of its competitive strategy. It also helps us attract professional people to work for the firm." But Crown has worked hard not to become a "company town" in the unattractive sense of the phrase. The Crown logo appears so discreetly on so few doors that a stranger passing through town would have trouble knowing that Crown even exists.

In the early years, recalls accountant Lois Moeller, some of New Bremen's citizens resented the Dickes buying up land; one or two even suggested sarcastically that they call it Dickeville. "Nobody really knew what they were doing for the town, and a lot of people got the wrong idea, because a lot of what they did, they did anonymously," says Lois. "But most people now really look

Honoring tradition

Crown's international headquarters, the Carl H. Dicke building, erected in 1994, combines old-world and airy new-world styles. Faced in stone and roofed in green slate, with leaded glass windows and oak paneling, the flooring is virgin Ohio timber harvested from old barns. "It was influenced" writes Jane Ware in Ohio Magazine, *"by numerous pictures of old meeting houses that Jim II took on business trips to Germany." "It was intended," says Jim II, "as a building that might have been done in New Bremen 120 years ago. In fact, visitors sometimes ask us about the restoration, not realizing it is a new building."*

up to them and feel they have done wonders for the town, and that you can see it."

"If it weren't for Crown doing what they've done with the downtown, we would have been another dying downtown that looked run down," says Kathy Topp, manager of the Crown Store. "We were starting to look that way." The Ohio Historical Society recognized Crown's effort with a Preservation Merit Award. "Remodeling and restoring these older buildings," says Jim II, "allows us to create a positive impression of the whole community and of Crown. We like to hear customers remark, 'Crown makes a quality product in a wonderful farming community.'"

Years ago, Crown purchased its own jet so it could fly customers in to see how the lift trucks were manufactured. "When we invite them to come, they invariably say 'No,'" says one of the salesmen. "If we say to them, 'Do you have time to get to the airport? We'll fly you in and have you back for dinner,' then the customers say 'Yes.' In hard times, many companies cut back on things like corporate aircraft, which stockholder groups consider luxuries. But Crown has used its aircraft to sustain its business.

There are no hotels in New Bremen, so over the years Crown has gradually developed guest quarters by restoring a farmhouse, a railroad car, a train depot, and a Queen Anne home. On one occasion, the guest houses influenced a big contract with a Cincinnati firm. The Access Corporation was building a piece of equipment that mechanically selected IBM data cards (in the early stages of computer automation). Access had a sales organization but didn't know how to manufacture the product so, in a meeting in Cincinnati, they asked if Crown might be interested in doing so. But Access executives expressed doubts about whether Crown could handle production: "You're a pretty small company in a pretty small town," they said. So Crown invited them to New Bremen and put them up in the guest house, where they spent the evening together, eating and socializing. The next morning Verlin Hirschfeld picked them up and asked "Well, did you guys sleep well?"

"Yes, we slept real well," was the reply, "and since we woke up we have decided to do business with you."

"I've always said, if we had put them in a hotel they would have gone away and said, 'Let's talk about it in a month,' but they were impressed," says Verlin. "It worked because it was a small town, and because we spent the day and evening with them, then turned the guest house over to them, saying they had the run of the place. It's a heck of a lot different than putting people in a motel down the road, where they feel they are just in another public place. Treating people like real people has paid off every day of the week. It's the same principle as Jim talking to people on the factory floor. People like to be treated like people." ❍

[1] *Neil Armstrong, the first man to land on the moon, was born and raised in Wapakoneta, Ohio, the county seat. Armstrong and Jim Sr.'s good friend, Dr. Don Schwieterman, of Maria Stein, a small town southwest of New Bremen, were made Eagle Scouts on the same day.*

Home base *Much of Crown's manufacturing is concentrated in this New Bremen complex. As the lift trucks evolved and manufacturing became more complicated, Crown grew to accommodate production needs. Opposite, Bob Marshall and Mose Zimpfer.*

and Improving Expanding Production

11.

Asked why he felt such loyalty to Crown, Mose Zimpfer explains, "They gave me the chance that nobody else would probably have ever given me. Without a high school diploma or a college degree, I would never have had the opportunities I've had at Crown. They knew my father was a welder and that I had some background and they took a chance on me." Mose is a good example not only of how Crown built company loyalty but of how it encouraged production workers to figure out ways to improve Crown's production technology. Mose had started at Crown in 1962 as one of the first production welders on the lift trucks. (Welding is like putting in the seams that hold metal parts together.) Mose's enthusiasm for the company began to grow the first day he met Jim Sr. Jim walked up to his work station and said, "Who are you?"

"I'm Mose Zimpfer. They tell me that you're Jim Dicke. Is that correct?"

"Yes, that's correct."

"You don't care if I call you Jim then, right? You can call me Mose because nobody knows me by my right name." (Ronald is his right name.) Some time passed and when Jim stopped by Mose's work station again, Mose said, "I'd like to show you a new way of welding."

"How's that?" said Jim.

"It's called microwire." (Continuous arc welding was called wire welding because a roll of welding wire on a spool or coil is used, in contrast with the stick welding rods used in conventional arc welding.) "I'd like to show you what I can do with it."

Jim told Mose to arrange to get a continuous arc welder in the plant so he could see Mose work with it, so Mose arranged for a firm to bring it in — the salesman happened to be Don Wedge, a referee for the National Football League — and Mose ran it for two months. Jim came by one day and asked Mose, "What if I told you you had to get rid of that?"

"I'd quit," said Mose.

"Why would you quit?"

"Because it would mean you want to work in the horse and buggy days and I'm not a horse and buggy driver. I'm here to do things the newest way that I can do them, the fastest."

"You order two of those," said Jim, and Mose did. "That started the wire welding here at Crown," says Mose.

Once Crown got into making lift trucks, it was critical that key truck components be consistently high in quality, because the trucks had to be safe, durable, easy to maintain, and easy to use. But going to a high level of production is expensive. In the early years of developing the lift truck, contract work supported that cost and gave Crown experience in precision manufacturing.

With lift trucks you can't afford to have a bad weld because a welder has a bad day; you need consistent production. So in 1977 Crown bought its first robot, for welding. When they first got into robots, Crown tried to work with Hobart in Troy. Hobart's vice-president of engineering told Arnie Heitkamp, "You're nuts. What you guys want to do can't be done." So Arnie contacted a company in Japan, which said it would be no problem. The Japanese firm, Snin Miewa, shipped parts to a distributor in Detroit. Robot welding was a new concept even to the dealer in Detroit, so the president of the Japanese firm flew into Detroit and, using a robot, welded some pump bodies together for Crown to use in manufacturing pallet trucks. The robot did a beautiful job, taking three or four minutes to do what it usually took about fifteen minutes to do at Crown. Crown bought two of the machines.

"I didn't know anything about robot welding till Jim had them brought in here," says Mose Zimpfer. In the somewhat comical period that followed, a Japanese man who spoke no English had to communicate with production workers like Mose, who had to draw pictures to show the Japanese man what they wanted done. Finally Mose asked his boss, Arnie Heitkamp, to send the Japanese man home; he would figure out how to run the thing himself. Mose had taught himself electronics, and Crown now encouraged him to learn enough about computers so that he could program the robot to do what it was supposed to do. "It was a new process and it was unheard of around this part of the country," says Mose. Crown was one of the first firms in the country to have robot welders, and Mose — who had never finished high school — was soon conducting demonstrations for firms from all over the country, including Hobart, which under normal circumstances should probably have been conducting demos for Crown. Over the years Crown bought eleven robots. Eventually Hobart got into robots and sold some to Crown.

Production was full of challenges, because there were always new products to make and new technologies with which to make them. One of the hardest problems toolmaster Wilbert Will ever faced was figuring out how to get a robot to put pins in the pinion in the third gear of the antenna rotator. Crown had to send a man down to IBM's Florida plant to learn how to operate the robot (this time, an IBM robot) so he would know how to program the computer. Wilbert's tool design department had to develop a head for the robot that could pick up three different parts: a pinion, two pins, and a gear. The robot had to pick a pinion and then the pins out of a vibrating bowl and insert them in the gear.

"My job was different every day," says Wilbert. "Always building something new. It was interesting when it worked. We built a lot of big welding fixtures, dozens of them. Big fixtures, twice as big as a desk. You weld it and rotate it and then you weld the bottom side." Along the way, as the lift trucks got fancier, Crown employees also had to learn hydraulics and how to lift the masts of the huge lift trucks straight up. They had to learn not only how to operate them, but how to manufacture them as well.

People power

Crown produces most of its own parts and Crown's production workers are skilled at precision manufacturing and encouraged to suggest ways to improve production.

One of Crown's guiding principles over the years has been to get a part from a supplier if the supplier was reliable and the quality was good; if not, Crown would make it themselves so they could control quality and make sure they got delivery on time. Now that they buy and produce parts all over the world, currency exchanges also affect their decisions: Sometimes it's good to buy from Mexico or Germany and sometimes it's not. They also factor in capacities at the different Crown plants. They bought many parts from their German plant until their lift truck business in Germany developed, partly because Germany had the capacity and New Bremen could delay expanding, and partly because Crown knew they would do a good job.

The Mexican plant builds all the forged forks and some of the weldment sub-assemblies for the other factories. In addition to building powered high-lift trucks, the New Bremen plants build major components for the other factories. Crown uses outside vendors for controllers, pumps, valves, tires, castings, some gearing, and some motors.

Many of Crown's manufacturing activities are concentrated in the main New Bremen complex. Across the street from the engineering center are several component fabrication facilities and assembly plants. There Crown does everything from assembly and testing of printed circuit boards to chassis fabrication and final assembly.

Not only has Crown expanded its manufacturing operations to include overseas operations but it now has U.S. plants outside of New Bremen. In New Knoxville, Ohio, Crown built a 45,000-square-foot plant to design, develop, and manufacture *dc* (direct current) motors and injection-molded plastic components. All powered pallet trucks and tow tractors are manufactured in the firm's Kinston, North Carolina plant (Kinston Neuse Corporation, or KNC — run until recently by Jim Sr.'s younger son, Dane Dicke). Crown has expanded partly to tap different labor pools and partly to divide operations off so that each plant can concentrate on a certain product. If every product were a part of the New Bremen operation, 20,000 to 30,000 parts would have to be tracked on the New Bremen computer. Instead, the pallet truck, for example (described on page 85), is isolated at one site with one management in Kinston. With one team giving it full attention, questions are more likely to be voiced — such as, "Why aren't we doing this?" — and change is much easier. Crown has also recently established a plant in Greencastle, Indiana, where several counterbalanced models are manufactured.

At the time of the case study, Crown used a flexible mode of production, not an assembly line mode. Crown had low volumes and a great variety of products, so automated assembly operations weren't practical. Components were manufactured in batches — usually about two weeks' worth at a time — and transported by lift truck to the assembly plant, where trucks were typically assembled in batches. The case study explains: "Each assembly operation (such as installing the seats) took place more or less at the same time for all trucks in a given batch. After a given assembly operation was complete, another operation would begin in the same general work area. At certain points the batch of trucks would be moved to a new work area, but in general, the trucks (especially the largest ones) were moved very little during assembly operations. The sequence of assembly was organized so that standard subcomponents and sub-assemblies were attached first and customizing options were put on later." One of the chief advantages of this production process (over the assembly line) was its flexibility: You could alter the sequence of products being assembled. If optional parts for a particular order were not ready, for example, the truck or trucks could be set aside without holding up production of other trucks until those parts arrive on line.

In 1969, a tilt table was built in Plant 3 to test lift trucks to see where their center of gravity was and how far you could tilt them before they would start to tip over. Crown uses the tip-over data to determine the load-capacity ratings for its trucks, using guidelines established by the American National Standards Institute.

"As the lift trucks evolved, the manufacturing process became more and more complicated," says Wilbert Will, "and we just automatically grew with it." The biggest fixture Wilbert designed, before he retired in 1990, was a floor-level tilt fixture about as big as a single-car garage. Crown builds its big turret stockpicker trucks lying down. The tilt fixture grabs the truck by the base and stands it up, tilting it from the bottom. To build the fixture, Wilbert had a hole dug four feet deep in the floor for the length and width of the fixture (roughly 10 feet by 20), built the lower portion of the tilt fixture into that hole, and covered the whole thing with steel plates to form a table, with an additional fixture above it. The tilt table is hinged, so that half of the fixture surface comes up to a vertical position; the truck is fastened to the half that tilts up. When the truck is raised to an upright position, the clamps are released and the truck is moved to another area for completion.

Crown's product design has always been geared to thoroughly eliminate any problems with product safety. "We used to spend most of our time worrying about getting additional business or orders," says Tom Bidwell. "Now we spend a lot of time with computer people and lawyers — and we make fork lift trucks! That doesn't make any sense to me. We also have to make sure our plants are in compliance with all manner of government regulations. We have a hazardous materials manager, a safety manager, a full-time medical director, and six to eight nurses. I spend more time now with lawyers and doctors than I do with salesmen and engineers." The only advantage of all the U.S. product liability problems is that it keeps most German or European manufacturers out of the U.S. market. "They get one taste of this market, pack up their bags and go home," says Bidwell. "It's much easier over there. They don't have to put up with that problem." ❍

Kelly's the name, call me Ned *Ned Kelly, above and at left on opposite page, ran Crown's Australian operation in the early days. A genial, teasing salesman trained at National Cash Register's sales school in Dayton, Ohio, he radiated charm and good humor, often playing at being the charlatan. He often talked with other salesmen about the "quest for the blue vase" — in other words, the perfect sale.*

Crown Goes International

12.

In 1965, Crown launched its first overseas operation in Australia. Crown would not have gone into Australia that early if Jim Dicke hadn't belonged to the Young Presidents' Organization, an organization for people who become president of a mid-size or larger firm before the age of forty. YPO expanded Jim's awareness of business possibilities, and encouragement from fellow YPOers gave Jim the courage to branch out that he might not otherwise have had so soon.

Tony Vanderstraaten, who joined Crown's sales team in 1968, says, "In 1965, it took plain guts for a small firm in the U.S. farmlands to go halfway across the world and act like a big company, to say to the world, 'We are here, we are going to show here, and this is our equipment — it is built in New Bremen, Ohio. You guys should buy something from us.' We always acted bigger than we really were. Most potential customers assumed Crown to be a substantial American company. We were also riding on the coattails of our importers. Most of them were much larger than Crown. Of course, where the product is made is not that important, providing the product is top-notch in design and the price is right."

"In many ways Crown's expansion overseas paralleled the development of its product line in lift trucks," says Tom Bidwell. "As we added products here, we continued to add distributors and as our line got broader it became more attractive and it was easier to get distributors. Then we started to look at the export business. We sold initially through export agents, with rather limited success, but that did give us some exposure overseas. Our first real overseas adventure was in Australia. It started out, I think, with two people who in turn hired people and did a nice job of marketing. The company grew very rapidly over there." Now, in addition to its marketing operation, Crown also has manufacturing facilities in Sydney, Australia: They manufacture a powered fork lift and stacker models for the Southeast Asian market. Sydney's basic line of lift trucks includes some products that aren't made in the United States, including the original lift table with which Crown started.

Crown had some lucky breaks along the way: getting the right people in Australia, coming up with the right product at the right time, being very lucky with currency exchange rates, and being able to take advantage of government guarantees available through the Export/Import Bank to create jobs for export. "Australia was a wonderful success story for Crown, beyond anyone's wildest dreams," says Tony Van der Straaten, "and looking back it is easy to understand why it all happened as it did. But thirty years ago, to follow this long-term business strategy was almost unheard of. It took a long time for Crown to establish itself as a major player on the worldwide lift truck field. None of this could have happened without the owners' commitment and top management's foresight."

"Our first overseas adventure was a tremendous success," says Tom Bidwell, "so we thought, 'Well, there's nothing to this; we'll just do more Australias.' It helped that the U.S. government was promoting exports at the time as a way of creating jobs in the United States. And Crown was lucky: It got in on the ground floor in Australia while the lift truck market was relatively undeveloped there."

In 1968, three years after launching its Australian operation, Crown bought an Irish company that made hand pallet trucks. They bought the Irish firm from its parent company, the German manufacturing firm Steinbock, which was having marketing problems. "We thought, here's our chance to get a good, acceptable product that's already on the market," says Verlin Hirschfeld. "The Irish government was happy to subsidize it because Ireland had high unemployment, and as long as they could find somebody who would offer jobs they were happy to work with us."

Crown thought the Irish company — based in Galway, on Ireland's west coast — would be an instant success because the Australian branch had been. Crown succeeded in Ireland but not instantly. "It took about ten years," says Tom Bidwell. "We went through about three sets of management before we sorted all of that out. When problems first arose, we went over there taking various U.S. advisors, thinking, 'We'll handle this.' We didn't realize that when you are in a foreign culture, the culture is different and the laws are different." The problems finally got settled and Crown got the right management team together. "After about ten years, Crown in Ireland became a successful operation," says Bidwell. "It was also an excellent learning experience."

The Irish plant produces hand pallet trucks that are shipped to the United States and finished in New Bremen. (Pallet trucks are useful for transporting pallets of goods. You slip the forks under a pallet, jack it up, and then walk it — pushing or pulling it — to where you want it to go. You pump the handle to raise the pallet and operate a release lever to lower it.) When a shipment comes in from Ireland, Crown employees in New Bremen (and now in the Dayton facility) unload, assemble, paint, and then ship the trucks out to fill orders. The Irish plant now also produces European versions of the pallet walker, the pallet counterbalance truck, the sitdown counterbalance truck, and the B models.

The same year that Crown bought the Irish firm, it opened a sales and service branch outside of London, in excellent facilities near Heathrow Airport.

In 1972, Crown sent Tony Van der Straaten, who had joined the firm in 1968, to Europe as European sales manager based in Brussels. "Our sales volume at that time was 'peanuts' compared to today," says Tony. "Twenty-five years ago our competitors in Europe barely realized Crown was a viable company. That has certainly changed today. In the mid-sixties Crown attended a number of truck shows in Europe, and through this exposure got in contact with a good number of material handling dealers in European countries. In those early years we were shipping finished trucks from New Bremen, Ohio, to Europe, because the exchange rate was very good and freight was lower than what you might think. Crown's technology and design were a bit ahead of the Europeans then (the gap is smaller now), and we sold thousands of lift trucks — mainly in Holland, Belgium, Spain, and England. In those days we sold better in England and the Benelux than in Germany and France. In those days, it was also not easy for a German importer to sell an American-built lift truck. From the 1960s through about 1981 our German importer was a company with a head office in Amsterdam, Geveke Intern Transport. Many European firms were glad to get a reasonably priced lift truck, and we offered a very good product for the money."

Crown's first try at a dealership in France was singularly unsuccessful. Crown stuck it out for a while, but the French dealership was a lesson on how things sometimes don't work out.

By the time Crown opened its Mexican plant, in Querétaro, Mexico, in 1973, it had learned a few things. "In Mexico, we started off very well because we opened with an excellent Mexican partner, German Campos," says Tom Bidwell. "We had learned from our previous experience in other countries to be aware of cultural differences, so we didn't make the same mistake twice. We more or less let Campos be the guiding light and stayed out of it except for supporting him."

Crown bought all the capital equipment in New Bremen and shipped it to the plant in Mexico. Campos, as managing director and co-owner, set up the entire plant without Crown's involvement. After the plant had been there for a while, Arnie Heitkamp went down to help them work out how to improve the plant layout and processes, how to stimulate production, and how to improve the throughput of parts being made for New Bremen and Kinston. The Mexican plant now makes all of the forged forks and some of the sub-assemblies used in lift trucks produced and assembled in other Crown factories. Crown employs about 140 people — all Mexican nationals — in the Mexican plant.

Mexico's big currency devaluation in 1995 did not hurt Crown significantly because from that particular factory Crown is selling mostly for export. Whether Mexico can support a market for lift trucks may be questionable but management is looking ahead, saying "Hey, there's the future. We better get in there before somebody else does."

In 1986, Crown bought a factory in Roding, Germany, located in a Bavarian forest near Munich. Germany is home to Crown's leading international competitor, Linde AG, and Munich is right in the middle of Crown's toughest foreign market to crack — the tight, close-knit market of Germany, England, and France. "You learn very quickly," says Tom Bidwell, "it's not that they don't like Americans, it's that they don't *need* you, so they don't want you. Germans are very nationalistic and take great pride in their products. If it's not made in Germany, they don't want it. They don't buy Italian refrigerators and they don't drink French beer. You have to realize going in that you are not going to be accepted with open arms. You are not going to be an overnight success, because they are cautious and leery of newcomers. You have to earn your way." The market for Crown's lift trucks may be bigger in Europe than in the United States, but in Europe Crown has competitors who have been there a long time. Crown has had to work its way up through the ranks.

Crown has learned that launching manufacturing enterprises in other countries does not, as Tom Bidwell puts it, "mean simply duplicating products and processes. We must be run by people who understand their own country and their own customers. We've tried hard to be an Irish company in Ireland, an Australian company in Australia, and now a European company in Europe, not an American company in Europe. Our model was the Opel car, which is successful in Germany and is considered a German car, although it is manufactured by an American company, General Motors. Most people in Europe don't even know that. That's how we wanted to be: We wanted to be a European company that happens to be owned by Americans. We've worked hard at that. We did not want to be an American company selling in Europe."

How do you try to be European in Europe? For one thing, says Bidwell, "you don't supply a German firm with American drawing paper, which is different from German drawing paper; if you're a European company in Germany, the drawing paper has to be German. You don't insist that all business be conducted in English, and you don't go around immediately calling everybody by their first names. Americans think that's the way to do business, but it's not done that way in Europe, especially in Germany: They think it's disrespectful. Crown also had to change the model number on the walkie counterbalance truck from WC to WB, because in Europe 'WC' signifies 'water closet,' or toilet, so the initials tended to evoke giggles rather than respect."

Only four or five of Crown's thousand employees in Europe in 1995 were Americans; the rest were Europeans. Crown's Munich facility can handle nineteen different languages. It is not uncommon for a European to speak at least three languages and some of them speak six. "As an international company the American Crown is going to have to work a little bit on handling more languages," says Bidwell. "You cannot be an international company and expect the rest of the world to speak English."

Crown's Roding plant manufactures hand pallet trucks and European versions of the power pallet trucks and rider equipment. "Our market share in Europe isn't close to what it is here," says Bidwell. But Crown has stabilized a market share that is strong for a U.S. firm, and the European market is bigger and faster growing than the U.S. market. In Europe, Crown has well-established competitors who do very well in their markets and don't do well in the United States. Most American companies have not done well in Europe. Crown plans to be the exception.

"In general," says Bidwell, "the Europeans are way ahead of most American manufacturers in terms of design, and they still build the quality." One problem in Europe is that they also have many sets of standards: German standards, French standards, European Community standards. "Any European manufacturer has to put up with a lot of bureaucracy in standards. It's worse there than it is here in terms of different standards for every country," says Bidwell.

In 1995, Crown had manufacturing operations in Australia, Ireland, Germany, and Mexico and had recently purchased Hamech Limited, a small but successful British manufacturer of lift trucks. European marketing, advertising, and distributed engineering operations were headquartered in Munich, Germany. And Crown had branch sales offices in Australia (Albury, Brisbane, Canberra, Melbourne, Newcastle, and Sydney), England (London and Manchester), France (Paris), Belgium (Brussels), Germany (Dortmund, Frankfurt, Hamburg, Hanover, Munich, Nurnberg, and Stuttgart), New Zealand (Auckland, Christchurch, Palmeston, and Wellington), and Singapore. ❍

Recognition counts *Company loyalty builds when people are credited for their ideas and appreciated for their contributions. Above, Bill Hambrick, working on the reach truck assembly. Opposite page: Maintenance duties for Bob Wade (center) often included watering plants and shoveling sidewalks.*

People, Not Machines

13.

Crown was lucky. It entered the lift truck industry in the narrow-aisle niche when land was relatively inexpensive and buildings could sprawl. Then property became more valuable and the cost of square footage increased, so firms began to build upward, using what would normally have been dead space. Levitz, one of Crown's first big customers, needed trucks that could take storage upward. They needed a stockpicker with height, one that could lift a sofa off a shelf 22 feet in the air. The way the industry is changing is only increasing the market for material handling equipment. Retail warehouses, for example, used to occupy only 20,000, 30,000, or 40,000 square feet of space; now there are 160,000-square-foot stores served by mammoth warehouses far larger than the stores. All of this requires material handling equipment.

Lack of capital (to buy the machine tools needed to expand faster) may have been Crown's biggest weakness over the years. Yet that very lack of capital kept Crown from growing too fast, forced gradual growth at a time when fast growth might have been the company's downfall. By the 1970s and the 1980s, however, Crown was expanding rapidly. That growth has kept things exciting at Crown. The employees like it that "things were always moving, never standing still."

Why has Crown done so well? People at Crown are apt to say, "The harder we worked, the luckier we got."

"Luck has something to do with it," says Verlin Hirschfeld. "You have to have some luck, don't you? We found a good niche in the lift truck business that Clark and the big guys had long passed over, and we had some good product design from RichardsonSmith of Columbus. We've had good, hard-working local people who put in a good day's work — basically, farm people who did their best, more than we expected them to do. We have that going for us, plus management was aggressive. They were go-getters and made things happen — *worked* to make things happen. There was always another great idea and the Dickes knew how to listen."

Jim took the long view and encouraged others to do the same, in every aspect of the business. He also expected people to pay attention to details. That combination of paying attention to details and taking the long view paid off handsomely in the 1970s, when the lift truck business started to take off.

Over the years Crown worked with a lot of banks and always maintained good relationships with them. "Jim Sr. was great with bankers," says Roger Bornhorst. "He has a flare about him. He knew the products, he had a good sense of selling the company, and he always knew what he wanted to have done before he went into a deal. He would tell them what he wanted first; they would not say no to him. He would never argue — nobody argued with Jim — but he never let it get that far. He's very social. Jim II is like his dad in negotiations with bankers. They both know what they want when they go into a meeting."

And that good relationship with banks has paid off. During the early development period for lift trucks, it was probably as important as the quality of Crown's manufacturing and marketing. In the 1970s, inflation was high, so many people felt it was crazy not to borrow money — because the dollar you borrowed today would be worth less tomorrow and hence would be easier to pay back. But interest rates were about 7 or 7¼ percent and starting to nudge up a bit, so Crown, anxious about the rising rates, locked in several loans at 9½ percent, a rate nobody really expected the market to reach. When national interest rates reached 18 percent, it was clear that Crown's foresight had paid off.

"Crown has always been fortunate in its banking relationships," says Jim II. "The Loramie Banking Company (now the Star Bank) in Fort Loramie, Ohio, was always a great supporter of Crown's, eager to help in every appropriate way. When we grew beyond their lending limits, it was they who took us to the Piqua National Bank in Piqua, Ohio, where we have had most of Crown's major accounts for many years. Crown also has a good relationship with the local New Bremen bank, which serves not only the company but many Crown employees. The Minster State Bank has also been great for Crown's people to work with. In the 1960s, we began to outgrow the Piqua Bank's lending limits, too. Typically, at that point, a firm like Crown would have begun dealing with a money center bank in Chicago or New York, Boston or Pittsburgh. Big-city Ohio banks in Cincinnati, Dayton, Cleveland, and Columbus had a reputation for being more insular and not interested in small, rural Ohio manufacturing companies. Crown was fortunate. Jim Dicke, Verlin Hirschfeld, and Tom Bidwell hit it off immediately with the people at Society Bank (now the Key Corp.), and that relationship has continued unabated for over thirty years."

Jim II

"Not everything in material handling has been invented," says Jim Dicke II, Crown's president since 1980. "The whole industry continues to grow slowly, but it is going to be a different industry."

"Our challenge in the last few years has been to transplant our local attitudes of trust and respect and our ability to work together to our other facilities around the world," says Tom Bidwell. "Not only our factories, but all phases of our global operation." At the same time, Crown has learned from its international operations that its U.S. operations must develop more depth in language skills and understanding of other cultures. Crown U.S.A. needs more people speaking German, Spanish, and the languages of the other countries they'll be dealing with.

"Jim II runs the company," says Verlin Hirschfeld, "but his dad is still a very big helper and always will be. He hasn't changed. He still has good ideas, and I'm sure he throws them out to Jim II and they either decide to run with them or not. I hope he hangs in there for a long time."

Crown is a real family company. Jim II and Janet's son, Jim III, is vice president of Human Resources. Their daughter, Jennifer, works at Crown during college breaks and has been through the company's sales training.

"The challenges of managing a closely held corporation are not that different from those a public corporation would face," says Jim II. "Candor, communication, and a belief that the company comes first are very important. The difference is that in a private company you spend Christmas with your stockholders. Of course, changes are always going on in private companies just as in public ones. To flourish, companies must keep reinventing themselves."

Crown is too big now for the fast walk through the plant that Jim used to take, says Tom Bidwell. "A walk through our factory today will take four hours; that's half the day and that's a long walk, so you can't do it. We have to fight the big company syndrome, but it can't be the way it was. There was a lot of satisfaction when we were smaller. You could see things grow — I mean, you could see it physically. You saw the smiles on people's faces. That goes away after a while. You have this big elephant now that you want to move in some direction and you're not sure where to nudge it and it's hard to see the result when you do."

Still, Crown does not have the problem with turnover that other companies have. It is loaded with employees who have worked only for Crown, who came when they were young and never left, and who are generally satisfied with the way things are. Every year, several employees are inducted into the Twenty-Five Year Club; of the 171 members in 1995, only 53 were retired and only seven had died. People who come to work for Crown seem to stay forever.

It's not only that they stay for years, but they seem to enjoy coming to work. "Sunday was my worst day," says Verlin Hirschfeld. "I was always thinking about what am I going to do on Monday. Of course, you didn't know what you were going to do on Monday anyway. You didn't really plan it, it planned you."

"How could anybody ever come to work and be bored and say they had nothing to do?" asks Arnie Heitkamp. "Because there is so much to do. You could educate yourself every day. When I started here, boy, there were guys that would just amaze me; they'd stand here and tell me this is die casting, this is sand casting, this is plaster casting and I thought, 'Well, I'm not going to last long here. I don't know what this guy is talking about.'"

Jim Sr. considers money about fifth in importance in terms of rewards to employees. Recognition is first. "People like to be able to communicate directly with people, to be appreciated for what they contribute. They like recognition — simple things like giving them credit for an idea. When you give a person credit for an idea, he will help you more enthusiastically. When you create an atmosphere, starting at the top, where people get credit and recognition, then you don't get the backbiting and politicking that goes on in other companies.

"People stay with companies if they are treated properly and are given opportunities to grow in the company," says Jim Sr. "To get and keep good people you have to make opportunities available. If you start to go down as a company, what happens is your young good people are the first ones to leave. The older group will not leave because they don't have any other place to go and they have a certain feeling of security. Then the whole company starts to deteriorate because you don't have the people.

"A company has to keep growing," Jim continues. "You can't stand still — there is no such *thing* as standing still: You are going to either go up or go down, so you always have to go up. If you are not a growing organization, your people are not interested in being with you because they can't grow, they can't move up. And the people are the only thing that's important in the company. All the brick and mortar, all the machine tools, all the ideas, all your investments don't mean a thing if you don't have good people." ❍

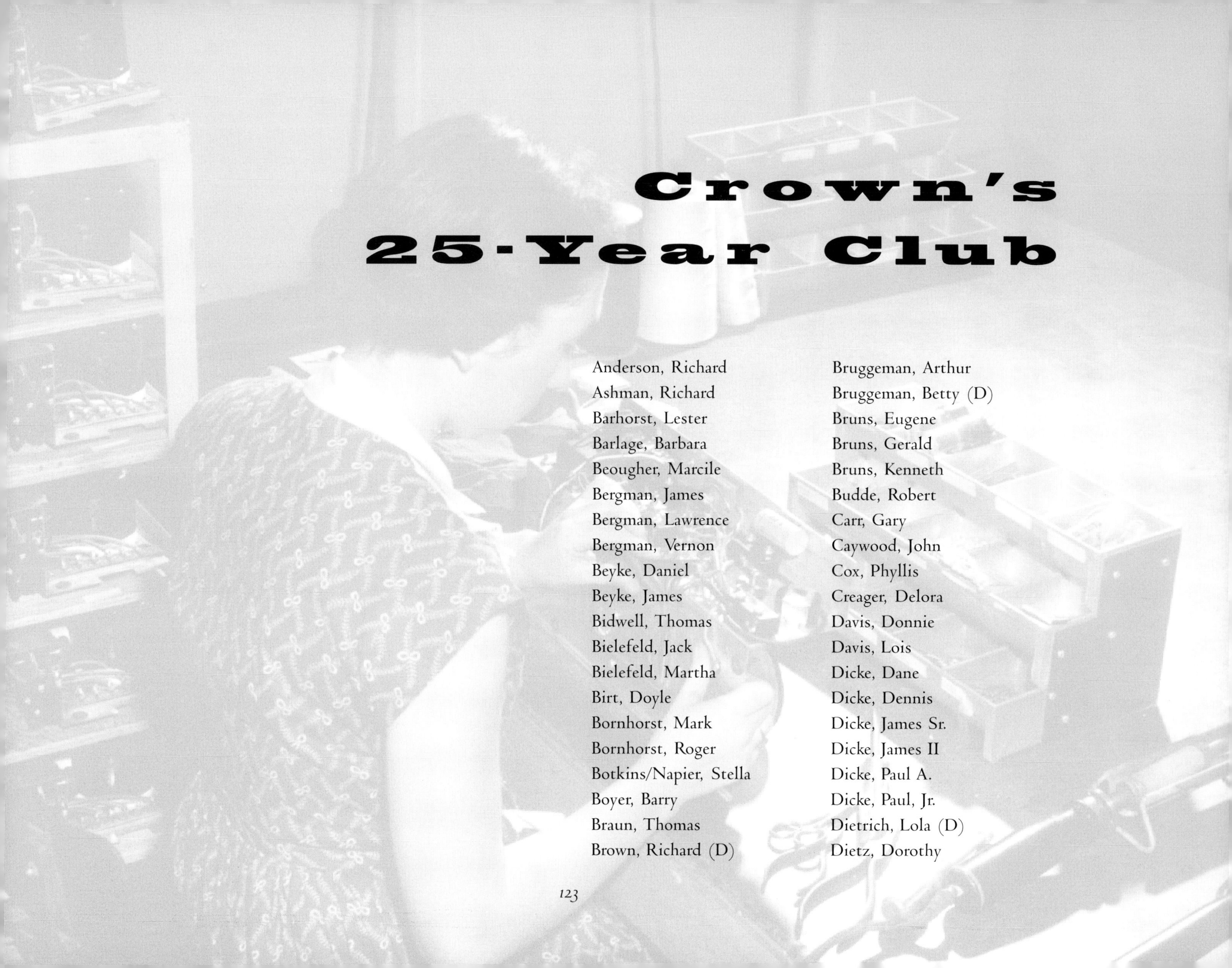

Crown's 25-Year Club

Anderson, Richard
Ashman, Richard
Barhorst, Lester
Barlage, Barbara
Beougher, Marcile
Bergman, James
Bergman, Lawrence
Bergman, Vernon
Beyke, Daniel
Beyke, James
Bidwell, Thomas
Bielefeld, Jack
Bielefeld, Martha
Birt, Doyle
Bornhorst, Mark
Bornhorst, Roger
Botkins/Napier, Stella
Boyer, Barry
Braun, Thomas
Brown, Richard (D)
Bruggeman, Arthur
Bruggeman, Betty (D)
Bruns, Eugene
Bruns, Gerald
Bruns, Kenneth
Budde, Robert
Carr, Gary
Caywood, John
Cox, Phyllis
Creager, Delora
Davis, Donnie
Davis, Lois
Dicke, Dane
Dicke, Dennis
Dicke, James Sr.
Dicke, James II
Dicke, Paul A.
Dicke, Paul, Jr.
Dietrich, Lola (D)
Dietz, Dorothy

Dobmeyer, Frederick
Doenges, Andrew
Droesch, Mary
Dunn, Ronald
Egbert, Lawrence
Enicks, Constance
Evers, Michael
Evers, Willard
Falkner, Richard
Fledderjohann, Max
Frank, Everett
Freewalt, Barbara
Galbreath, David
Garman, Tom
Garringer, Frances
Garringer, William (D)
Gehret, Edward
Gels, William
Gibboney, David
Gilberg, Juanita
Gilberg, Ronald
Grieshop, Thomas
Gruebmeyer, Wayne
Haley, Russell
Harshbarger, William
Hartings, Arlene
Hartings, Vernon
Heckman, Jeffrey
Hegemier, Duane

Heilers, Jim
Heitkamp, Arnold
Heitkamp, Eugene
Heitkamp, Patrick
Henegar, Shirley
Hespe, Mary
Hess, Richard
Hirschfeld, Craig
Hirschfeld, Verlin
Hoenie, Marvin
Homan, Leon
Hoskins, Jean
Hoyng, Thomas
Irish/Jacquay, Judy
Jameson, Ernie (D)
Johns, Gerald
Jung, LaRose (D)
Kemper, Jerome
Kessen, James
Kettler, Donald
Kinsella, Thomas
Klingler, Dennis
Knapke, Thomas
Koenig, Roger
Koeper, John
Koesters, Joseph
Koverman, Bernard
Koverman, Carl
Kramer, Doris

Kramer, Robert
Kremer, Lester
Kremer, Virgil
Krieg, Harold
Kruse, Dorothy
Lange, Carl
Lawler, Jack
Leasor, Paul
Lengerich, Bernard
Liesner, Frank
Link, Daniel
Link, Ivo
Link, Larry
Marshall, Robert
Martin, Agnes
McEldowney, Audrey
McGlinch, Kelley
McGlothen, Roberta
McIntire, Benjamin
McMaster, Duane
McMaster, Nancy
McMaster, Ray
McMurray, Steve
Meckstroth, James
Mertz, Bernard
Miller, Donald
Miller, Mark
Moeller, Adrian (D)
Moeller, Lois Gensler

Moeller, Mark
Moeller, Ned
Moran, James
Mouk, Robert
Mueller, Frederick
Nerderman, James
Nieberding, Robert
Niemeyer, Donna
O'Brien, Thomas
Oen, Gregory
Opperman, Betty
Osterholt, Leo, Jr.
Otting, Robert
Pulskamp, Gerald
Puthoff, Robert
Rempe, Pauline (D)
Rengers, Philip
Romie, Richard
Rose, Alfred
Ruley, Norman
St. Myer, Gary
Schaefer, Dale
Scheer, Donald
Schlarman, Michael
Schlueter, Irene
Schmehl, Mary
Schmitmeyer, Dale
Schmitmeyer, Roger
Schmitmeyer, Ronald

Schnippel, Joseph
Schoby, Rick
Schott, Kenneth
Schrage, Erna (D)
Schroer, Mildred
Schroer, Stanley
Schroer, Wayne
Schulte, Leo
Schulze, Fred
Schulze, Lavern
Schulze, Richard
Schwartz, Eugene
Schwartz, Ronald
Schwieterman, Alvin (D)
Seitz, Anthony
Sextro, Edward
Shaffer, Lola
Short, Earl (D)
Slife, Jackie
Smith, Arnold
Smith, Nelva
Spicer, Michael
Spradlin, Cleo (D)
Spradlin, Ivina
Stammen, Harold
Steinlage, James
Streight, James
Stucke, Delbert
Teeters, Richard
Thaman, Arthur
Thieman, Gary
Thornsberry, James
Tontrup, Edna
Tontrup, Jim
Topp, Kathleen
Trempert, Ralph
Triplett, Steven
Uetrecht, James
Vanderhorst, Daryl
Van der Straaten, Anthony
Vannette, Douglas
Vondenhuevel, Joanne
Voress, George, Jr.
Voress, George, Sr.
Wade, Robert
Watercutter, Richard
Wehmeyer, Floribel
Weimert, Donald
Wellemeyer, Fred
Wierwille, Robert
Will, Michael
Will, Theresa
Will, Wilbert
Williams, William
Wilson, Frances
Wilson, Robert (D)
Young, William
Ziegenbusch, Thomas
Zimpfer, Ronald
Zwiebel, Charles

Over the years, Crown lift trucks have received more than 30 awards for design excellence. Crown's first design awards, between 1965 and 1975, came from the American Iron & Steel Institute, which recognized four pieces of Crown Equipment with awards for industrial products and equipment design in steel. Between 1989 and 1997, several Crown trucks received the Industrial Design Excellence Award from the Industrial Designers Society of America, sponsored by *Business Week*. In 1990 IDSA named Crown's rider reach truck the Design of the Decade. That same year, *Business Week* named the turret stockpicker (TSP Series) "The Best of 1989." Between 1969 and 1997, Crown received the Industry Form Award, Outstanding Design of an Industrial Product, a dozen times at Hannover Messe, the industry's main international trade fair. Between 1972 and 1995, Crown trucks received nine Design Distinction Awards from *ID*, the magazine of international design.

Pictured here are the members of the design team for the FC Series counterbalanced rider truck, which in 1991 received three major industry awards. Front row, left to right, Keith Kresge (RS), Greg Breiding (RS), Mike Will, Dave Wilker, Carl Pohl, Gary Topp, Gary St. Myer, Henry Dickman, Gene Bruns, Nathan Blake, Frank Wilgus (RS), Harold Heitkamp. Others, left to right, Harold Stammen, Steve Billger, Dan Sherman, Jeff Vogel, Tom Linn, Lou Gehret, Ken Dicke, Larry Niemeyer, Inars Jurians (RS), Bob Roetgerman, Walt Conley, Bob Mervar (RS), Paul Stork, Jack Tarasovitch, Bob Drobeck, Dave Smith.

Timeline

1927
Allen, Carl, and Oscar Dicke launch Pioneer Heat Regulator Co. in Bloomfield, New Jersey.

1931
Pioneer moves to New Bremen.

1932
Master Electric buys Pioneer.

1945
Carl and Allen Dicke launch Crown Controls in the Rabe Hardware Store, 124 S. Washington Street, New Bremen, Ohio (Plant 1). They hire first employees: Lois Moeller (part-time)

and Bill Havemann. James F. Dicke returns from military service in November, goes on payroll (such as it is) in January 1946.

1947

Crown incorporates (March 17), purchases heat regulator division back from Master Electric, builds first addition (to north side of Plant 1), begins making own damper motor, produces first Ash Pit Spray Kit.

1949

Heat regulators become obsolete as the nation converts to gas. Crown ventures into a new field, television antenna rotators, with the Crown Tenn-A-Liner.

1950

Carl and Jim buy out Allen Dicke's interest in Crown; Allen takes over Carl's farm interests. With United States engaged in Korean "police action," Crown begins manufacturing parts for the military; government contracts make it possible for firm to begin doing precision machining. Meanwhile, firm shifts to better quality drive housing on the antenna rotator.

1951

Launch of Contract Services Division. First contracts: a phase shifter for radar, and turn and bank indicators for military aircraft. First addition to north side of Plant 1 extended to front sidewalk. Model CAR C antenna rotator introduced.

1952

Carl H. Dicke dies, December 18.

1953

To increase floor space, Jim purchases Arcade Building (Plant 2), builds southern addition to Plant 1 for shipping and receiving. Firm has 68 employees. First die-cast housing on rotator.

1954

Harold Stammen and Ken Griewe work in the crawl space area beneath Plant 1, welding parts for the radar phase shifter, as part of contract work for U.S. military. Crown introduces line of novelty products: master saw drill, arrowhead, armor cable cutter, adding ice stoppers in 1955.

1955

Crown introduces Model D antenna rotator and launches Canadian rotator operation. In Manheim, Pennsylvania, the Hershey Machine & Foundry Company assumes name of the parent organization, Fuller Company, Manheim Plant. Sales of motor stokers and Hershey oil burners, formerly handled by the Motorstoker division, to be handled by a separate company known as Hershey Motor Stoker Inc., which Jim Sr. will run. Crown hires engineer Tom Bidwell, who will play significant role in development of lift truck.

1956

"Utilitongs" added to list of Crown novelties. Firm gets first government contract for repair services. Although heat regulators are no longer a factor, Crown makes modulating control for schools. Introduce first Crown lift trucks, models LT500, LT1000, and FL1000.

1957
Channel Master makes offer Jim Sr. can't refuse: large-scale distribution of antenna rotator. Crown speeds up production, grows—has 117 employees when Bill Havemann (first fulltime employee) dies. Crown produces customized lift trucks for various industries (funeral, steel, postal, and others). Income from contract work subsidizes lift truck development.

1958
National recession.

1959
Channel Master Corporation takes over distribution and marketing of antenna rotator; Crown retains patent, continues manufacturing, discontinues Tenn-A-Liner name for antenna rotators. When Crown stops producing one-of-a-kind lift trucks and introduces standard line of H and B series (hand- and battery-operated), it truly enters material-handling industry. Annual lift truck sales: $50,000. Crown issues catalog showing several models of E-Z Lift line (plus line of Phoenix wheels & casters).

1960
Deane Richardson talks Crown into hiring fledgling design firm RichardsonSmith (RS) as outside designer of lift trucks. Crown shuts down heat regulator assembly line.

1961
Crown buys old Excello Building on west side of town (Plant 3). Engineering department split into Contract Services and Lift Trucks. Jim Dicke Sr. joins Young Presidents' Organization.

1962
Hand-controlled powered walkie pallet truck (W227 series) introduced, first design by RichardsonSmith (RS), which in 1965 wins Crown's first design award, from the Industrial Designers Institute. Add 20,000 sq ft to Excello Building, with 211 employees on payroll.

1963
First contract with Baldwin Pianos. Contract Services division launched, combining contract manufacturing and Gentile Repair Services. Last shipment of parts for heat regulator. Canadian operations moved to Crown's own plant in St. Thomas (26 employees), to handle manufacturing of antenna rotators and import of radios and record players. Crown adds space to Plant 3, has 244 employees (146 men, 98 women) on payroll.

1964
First contract with IBM (for part for Dictabelt).

1965
Design awards won for tow tractor/personnel carrier and pallet truck. Launch Australian operation, expand Plant 3 for lift truck production. Make forged-steel forks standard on 1500-lb capacity stackers with adjustable forks; start using hot-rolled steel mast sections on stackers of 2000 lbs and all telescopic models.

1966
Sears Roebuck Co. of Seattle orders 13 intermediate-duty walkie counterbalanced stackers, largest single order received for this model. Plant 3 gets large addition; Crown has 400 employees. Crown Australia launched in Sydney (Ned Kelly, general manager), to manufacture H, B, and W series (walkie stackers); sales & service branch opened in Melbourne, Australia.

1967
Series 20W and 30W heavy-duty walkie stackers go into production.

1968
"Shuttle" hand pallet truck introduced. Produce first SP and SPC stockpickers, WA heavy-duty articulated walkie stacker, and 3-wheel sitdown counterbalanced fork trucks, models Piccolift and Triolift. Purchase plant in Galway, Ireland, from German firm, Steinbock Ltd.; open sales & service branches in Chicago and London (near Heathrow Airport). Booz·Allen incorrectly predicts end of market for antenna rotators (still selling steadily in 1995)

1969
Stockpicker and walkie stacker win industry awards. Introduce heavier-duty WC walkie counterbalanced model (later changed to WB). Introduce counterbalanced stockpickers. Introduce complete line of lift trucks to UK at Crown London branch. Add more space to Plant 3 (now more than 116,000 sq ft); install test ramps and tilt table near Plant 3.

1970s
Ohio copes with period of economic readustment. Dayton, long a center for manufacturing hard goods (with six General Motors plants), loses 35,000 manufacturing jobs as Ohio and Michigan jockey for the nation's highest unemployment rates. Eventually the manufacturing base stabilizes, but Dayton diversifies, developing a mix of manufacturing, technology, and services. Wright-Patterson Air Force Base's expansion helps turn Dayton into Midwest's high-tech capital. With double-digit inflation and high unemployment in Dayton, it makes economic sense for Crown to expand.

1970
Jim Sr., Lois Moeller, and Wilbert Will become first members

of Crown's 25-year Club. Levitz (furniture discounter) places landmark order for 67 stockpickers, which gets momentum going on lift truck sales. Add refinements and new capacities on several truck models (including nine new models —with various fork widths and lengths—in shuttle hand pallet truck line). Install new mechanical brake on walkie stackers, produce first WPT heavy-duty walkie pallet trucks. Crown joins Industrial Truck Association; conducts first in-plant service clinic, starts sales training tape program; acquires NCR Century 100 computer. Sydney branch expands and construction begins on new plant in Melbourne. Worldwide, Crown has 150 lift truck dealers (80 in U.S.A., 70 overseas). Crown stops servicing heat regulators.

1971

Deliver Levitz stockpickers. RichardsonSmith releases first mock-up of three-wheeled sitdown counterbalanced truck (SC series), to be introduced after the RC series. Introduce walkie reach truck, end control and center control pallet trucks (PTE/PTC); successfully test new 22-foot high-lift stockpicker. Complete construction on plant in Melbourne, Australia, and expand sales, service, and storage branch in suburban Melbourne. Expand U.S. sales clinic to four days. "New difference" corporate brochure wins top honors from Printing Industry of America. Advertising theme for year: "Take On More With Crown."

1972

Introduce rider stand-up end-control counterbalanced lift truck (RC series)—the innovative side-stance design—which wins several design awards. Introduce tow tractors and sell first TWR to Michigan Food Warehouse. Break ground for Plant 4; remodel office front, Plant 2. Begin construction on new Sydney, Australia, plant, to manufacture complete Crown line. Introduce employee newsletter, "Crown News."

1973

Crown wins several design awards, for the walkie pallet truck, walkie stacker, walkie rider tow tractor, stand-up counterbalanced rider. Introduce powered retractable overhead guard as option on Series RC; first four-point suspension stockpicker. Complete Plant 4; open plant in Querétaro, Mexico, with partner German Campos. Issue Crown employee manual, with slide presentation. Open sales/service branch in Brisbane, Australia.

1974

Lift truck development, at break-even point, begins to take off; Crown gets out of government contract work, begins expanding at rapid pace. Introduce the sit-down counterbalanced truck (SC series) and tow tractor with "T" handle.

1975

Crown maintains leadership in stockpicker market by introducing expanded line of stockpickers. Sit-down counterbalanced rider (SC series) receives design award. Walkie counterbalanced stacker (WC series) becomes WB series (to avoid jokes in Europe about water closets); W series (intermediate duty walkie stackers) becomes M series.

1976
Complete first addition to Plant 4; in test lab, install adjustable ramp for testing high-lift trucks.

1977
Buy first robot for welding. Introduce new Series PTH hand pallet truck, to replace PT pallet truck.

1978
Crown producing 300 lift trucks a month. Make triple-stage mast available for SC and RC, improving visibility. Introduce slip sheet rider to the food industry. Complete Crown's data center, under direction of Dane Dicke.

1979
Build Plant 5 for steel storage. Produce first rider reach truck (RR series) for use in narrow aisles of warehouses and distribution centers.

1980
Rider reach truck (RR Series) is major success, wins awards immediately and, in 1990, ISDA's Design of the Decade award. Adds RD and RS models. Crown completes engineering building, has 1,195 employees. Jim II becomes president.

1981
Build second addition to Plant 5. Introduce GPW walkie pallet truck series.

1983
Convert downtown doctor's office to Crown Training Center.

1984
Introduce enormous man-up turret sideloader (TS), one of most advanced trucks of its kind in the world (result of research started in 1972). Build Plant 7.

1985
Control handle on walkie truck wins design award from *ID*, Magazine of International Design. Launch Crown medical department and dispensary

1986
During recession, go to seven-hour day, with some temporary layoffs. Launch German operations, purchasing a factory in Roding, Germany. Install rotating tilt table in test lab—only one in the lift truck industry. Move Advertising and Print Shop into two-story renovated building in downtown New Bremen. Add expanded "control room" to stockpicker, for additional operator comfort and productivity. (Operator area now 27 x 42 inch platform, which wins design award next year from *ID*,

Magazine of International Design.)

1987
Add heated cab option to TS series. Complete Building Maintenance building. Expand corporate offices into two adjacent buildings in downtown New Bremen. Purchase White Mountain for truck demos.

1988
Introduce elegant turret stockpicker (TSP), which captures several design awards. Finish addition to Engineering Building. Crown Controls becomes Crown Equipment Incorporated.

1989
Turret stockpicker (TSP Series) receives four major design awards; new European center-control pallet truck and walkie pallet truck also receive awards. Continue expansion, purchasing video rental store for office personnel, selling White Mountain, purchasing the old Opera House from American Legion, purchasing Stamm Building for Legal Department, and building next to Stamm Building for Design Group.

1990
Rider Reach Truck wins Design of the Decade Award from Industrial Designers Society of America (IDSA); *Business Week* names turret stockpicker (TSP Series) The Best of 1989. Crown introduces four-wheel counterbalanced rider truck. Plant 5 doubles in size; Crown buys land to add parking lot north of Plant 3; starts building motor plant in New Knoxville. Firm has 2,442 employees. Era of contract work ends, as Crown completes last contract job for IBM.

1991
Counterbalanced rider (FC Series) receives three major industry awards.

1993
Crown wins 1993 Beacon Award for Integrated Corporate Communications, sponsored by *Fortune* and the American Center for Design. Introduce ESR side-sit reach truck for European market.

1994
Crown has 3,186 employees. Introduce WE, WS European walkie stacker lift truck series. New PE3000 receives Industrial Design Excellence Award, Gold, from IDSA.

1995
More design awards received, for SP3000, WE2000, ESR 3000, and GPW1000 trucks. Crown producing 35,000 lift trucks a year. Add new parts warehouse (30,000 sq ft of high-bay rack), plus 18,000 sq feet on second level. Parts to be controlled by using Intermec bar coding and scanning equipment, in conjunction with a Rims Warehouse Management System, being updated by RF terminals. ❍

In people we trust *"The people are the only thing that's important in the company," says Jim Dicke Sr. "All the bricks and mortar, all the machine tools, all the ideas, all your investments don't mean a thing if you don't have good people." Shown here is production worker Charlie Heil, circa 1950.*

Index

Boldface indicates reference to caption or illustration.

About the Author

Pat McNees worked in book publishing several years (as an editor at Harper & Row and at Fawcett) before becoming a freelance writer and editor. Her writing, editing, and teaching for development organizations such as the World Bank has taken her as far afield as Burma and Lesotho and in the process of writing a history of the Young Presidents' Organization she has interviewed dozens of executives and entrepreneurs. She is the author of *An American Biography: An Industrialist Remembers the Twentieth Century*, and has edited several anthologies, including *Contemporary Latin American Short Stories* and *Dying, A Book of Comfort*. A member of the American Society of Journalists & Authors, the Authors Guild, the National Association of Science Writers, PEN, and the Society for Technical Communication, she lives in Bethesda, Maryland.